Revolutionizing Earth Observation - New Technologies and Insights

Edited by Rifaat M. Abdalla

Published in London, United Kingdom

Revolutionizing Earth Observation - New Technologies and Insights
http://dx.doi.org/10.5772/intechopen.1001487
Edited by Rifaat M. Abdalla

Contributors
Ahmed A.H. Siddig, Alexandre Flávio Assunção, Alexey Lyubushin, Anwarelsadat Elmahal, Böris Nörgaard, Carmen Marin, Cezar Scarlat, Cosmin Cara, Cosmin Udroin, Daniel R. Unger, David L. Kulhavy, Eltaib Ganwa, Felipe Augusto Nascimento de Jesus, I-Kuai Hung, Jamal Zamot, Laurentiu Nicola, Leonardo Santana de Oliveira Dias, Lindsay McBlane, Luis Ambrosio, Luis Iglesias, Michael A. Merchant, Mohammed Afkareen, Nasradeen A.H. Gadallah, Nistor Andrei, Pierre Houdmont, Reid Viegut, Sadri Haouet, Sophie Bontemps, Thomas Moore, Victoria Williams, Yanli Zhang, Yury Alex Polkanov, Zoltan Szantoi

First published in London, United Kingdom, 2024 by IntechOpen
IntechOpen is the global imprint of INTECHOPEN LIMITED, registered in England and Wales, registration number: 11086078, 167-169 Great Portland Street, London, W1W 5PF, United Kingdom

British Library Cataloguing-in-Publication Data
A catalogue record for this book is available from the British Library

Additional hard and PDF copies can be obtained from orders@intechopen.com

Revolutionizing Earth Observation - New Technologies and Insights
Edited by Rifaat M. Abdalla
p. cm.
Print ISBN 978-0-85466-710-9
Online ISBN 978-0-85466-709-3
eBook (PDF) ISBN 978-0-85466-711-6

For EU product safety concerns:
IN TECH d.o.o., Prolaz Marije Krucifikse Kozulić 3, 51000 Rijeka, Croatia,
info@intechopen.com or visit our website at intechopen.com.

Meet the editor

Dr. Rifaat Abdalla, PGeo, GISP, PMP, PMI-RMP, is an esteemed professor with great contributions to Geographic Information Systems (GIS) and remote sensing. His career spans academia, government, and industry, earning him accolades and awards for his work. Notably, he received the esteemed 2008 ASPRS Best Paper award, a prestigious recognition in geomatics, showcasing his outstanding contributions to the field. In 2022, Dr. Abdalla was honored with the Outstanding Leadership Award at the Education 2.0 Conference in Dubai, highlighting his exceptional leadership skills and mentorship in GIS and remote sensing. As a Professional Geoscientist (PGeo), Geographic Information Systems Professional (GISP), Project Management Professional (PMP), and Risk Management Professional (PMI-RMP), Dr. Abdalla exemplifies excellence in his field. His expertise and credentials testify to his impactful role in advancing GIS and remote sensing through pioneering research and teaching. Dr. Abdalla's work inspires future generations in the geospatial community.

Contents

Preface

Earth observation has evolved dramatically over the past few decades, transforming from a niche scientific endeavor to a cornerstone of our understanding and management of the planet's dynamic systems. The unprecedented technological advancements in this field have revolutionized the way we monitor, analyze, and respond to various environmental and societal challenges. As we stand on the brink of new horizons in Earth observation, it is imperative to merge these innovations and insights to guide future developments and applications.

Revolutionizing Earth Observation – New Technologies and Insights provides a comprehensive overview of the cutting-edge technologies and methodologies that are reshaping the landscape of Earth observation. This book is divided into two sections on technological advancements and practical applications.

The first section explores the latest advancements in Earth observation technologies. This part of the book examines new tools, methods, and systems that are enhancing our ability to monitor and understand the Earth's complex systems. From next-generation satellites and sensors to unmanned aerial vehicles (UAVs) and the integration of artificial intelligence (AI) and machine learning (ML), these innovations are driving the field forward. By harnessing these cutting-edge technologies, higher resolution imaging, more accurate data collection, and sophisticated analytical capabilities are achieved, enabling us to observe and interpret the Earth's processes with unprecedented detail and precision.

The second section highlights the practical applications of these advanced technologies across various domains. This part of the book presents real-world case studies demonstrating the impact and effectiveness of Earth observation technologies in addressing global challenges. Focusing on climate change monitoring and mitigation, agricultural innovations, and disaster response and management, the case studies illustrate how these technologies are being utilized to improve our understanding of environmental changes, enhance food security, and bolster disaster preparedness and response efforts. These applications showcase the transformative potential of Earth observation technologies in contributing to sustainable development and resilience.

As these insights and advancements are compiled and presented, the goal is to not only document the current state of Earth observation but also to illuminate the future challenges and trends that lie ahead. The evolving landscape of Earth observation will continue to be shaped by technological innovations, ethical considerations, and the need for international collaboration and policy development. By touching on these future challenges, this book provides a roadmap for researchers, practitioners, and policymakers to navigate the complexities and harness the full potential of Earth observation technologies.

Ultimately, *Revolutionizing Earth Observation – New Technologies and Insights* serves as a vital resource for anyone involved in the field of Earth observation. It celebrates the technological strides made, showcases the impactful applications of these technologies, and anticipates the future directions of this rapidly evolving discipline. This book is intended to inspire continued innovation and collaboration, driving forward collective efforts to observe, understand, and protect our planet.

Rifaat M. Abdalla
Associate Professor,
Sultan Qaboos University,
Muscat, Oman

Section 1

Innovative Technologies in Earth Observation

Chapter 1

Free Geospatial Software and Applications

Leonardo Santana de Oliveira Dias, Alexandre Flávio Assunção and Felipe Augusto Nascimento de Jesus

Abstract

This chapter examines the significant impact of Free and Open Source Software (FOSS) on the geospatial industry, tracing its development from the 1980s to its current widespread adoption. FOSS, which allows users the freedom to use, modify, and distribute software, has cultivated a varied ecosystem of projects, including geospatial applications under FOSS4G (Free and Open Source Software for Geospatial Applications). The Open Source Geospatial Foundation (OSGeo) supports the collaborative advancement and distribution of these technologies, leading to their global utilization. QGIS, a key component of FOSS4G, demonstrates the potential of open-source GIS software with its extensive plugin capabilities and support for various formats, such as the efficient GeoPackage format. Additionally, the Orfeo Toolbox (OTB) expands the reach to high-resolution satellite image processing, and the QuickMapService plugin in QGIS utilizes databases like OpenStreetMap (OSM) for enhanced geospatial analysis. The chapter presents case studies showcasing the practical use and effectiveness of FOSS4G tools in solving complex geospatial challenges, affirming their maturity and quality. Ultimately, the chapter highlights the importance of FOSS and FOSS4G in democratizing access to quality geospatial tools, enabling professionals to undertake sophisticated projects without licensing costs, with OSGeo's support ensuring their continual improvement and widespread adoption.

Keywords: geospatial science, FOSS, FOSS4G, spatial analysis, remote sensing

1. Introduction

Driven by initiatives that emerged in the 80s and 90s, the term FOSS (Free and Open Source Software) emerged and has grown exponentially since then. This term encompasses a wide range of projects, including operating systems, databases, text editors, applications for image and vector manipulation, programming languages, and development libraries, along with a multitude of applications that play a crucial role in how we interact with technology.

The 'Free' in FOSS refers to freedom, not price. Users have the freedom to control the program and use it as they wish. This contrasts sharply with proprietary software, where users do not have access to the source code and cannot modify, redistribute, or learn how the software was developed. In addition to freedom, there is a wide variety of software that can be downloaded and used without the need to pay for usage licenses. We would like to highlight some software that was useful in the preparation of this chapter, such as the text editor TexStudio (LaTeX), the image editor GIMP, the vector graphic editor InkScape, and the most popular GIS (Geographic Information System) open-source software today, QGIS!

FOSS initiatives have expanded into geospatial science, where they have become known as FOSS4G (Free and Open Source Software for Geospatial Applications). open-source geospatial software includes a wide variety of libraries, tools, applications, and development platforms, which are under various types of licenses provided by the OSI (Open Source Initiatives) [1] and FSF (Free Software Foundation).

For geospatial initiatives, there is a non-profit organization whose mission is to support and promote the collaborative development of open-source geospatial technologies and data, the OSGeo (Open Source Geospatial Foundation). The OSGeo acts as a public forum for sharing knowledge and ideas, contributing significantly to the FOSS4G universe. It plays an important role in promoting the use of open-source geospatial software and in forming a global, inclusive, and diverse community of contributors and users. Through its projects and initiatives, OSGeo is helping to shape the future of the geospatial sector.

The future is being shaped by the growing adoption of open-source geospatial tools, driving significant advancements in software projects through their extensive user base. The substantial uptake of FOSS4G tools is evidenced by the increasing number of published articles, showcasing case studies where complex issues have been addressed using open-source geospatial software.

In this chapter, we will present some free applications that can be used for the development of geospatial projects. These tools provide support to professionals working with geographic space, allowing them to benefit from a wide range of software and libraries, without the costs that commercial tools used in geospatial applications have.

We will also present case studies that offer insights and examples of the diversity and complexity of problems that FOSS4G tools can be the technological bases to solve. This will demonstrate the level of maturity and quality achieved by these tools today!!!

2. FOSS and FOSS4G conceptualization and evolution

The creativity, dynamism, and high-profile success stories of the Free and Open Source Software (FOSS) and FOSS for Geospatial applications (FOSS4G) movements are attracting increasing attention from end users, developers, businesses, governments, educators, and researchers around the world [2].

People everywhere are adopting various open-source distributions or participating in the general movement by contributing their own modifications [3], whether in the scope of the FOSS or FOSS4G initiatives. This involvement underscores the democratization of technology, fostering innovation and collaboration on a global scale.

FOSS and FOSS4G initiatives are revolutionizing GIS and remote sensing areas by empowering users to develop sophisticated projects. The open-source nature democratizes access, allowing diverse users to contribute to and benefit from a shared knowledge base. This promotes robust, flexible, and innovative solutions across a wide range of geospatial needs. From mapping and analysis to satellite imagery, FOSS and FOSS4G are driving global collaboration and innovation. Its success exemplifies the transformative power of open-source software in geospatial fields.

2.1 FOSS

FOSS has a long history, beginning with the origins of software itself, when the terms free software and open-source software were not yet defined [4]. The term FOSS means software whose license is compliant with the Free Software Foundation's list of four essential freedoms [5].

Free Software Foundation (FSF) and Open Source Initiative (OSI) has set up the definition for 'free software' and for 'open-source software', and according to the FSF, software can be labeled as free software if the associated license conditions fulfill the 'Free Software Definition' [6], which grants four freedoms:

- The freedom to run the program for any purpose;
- The freedom to study how the program works and adapt it to your needs;
- The freedom to redistribute copies so you can help your neighbors;
- The freedom to improve the program and to release your improvements to the public, so that the whole community benefits.

The FOSS movement has 20–40 years of history, having available a mature FOSS project for almost every software need and application; been considered crucial for the developing world because they enable emerging countries to develop their own technology instead of having to import it; provide a path to close the digital divide between rich and poor countries; create healthy competition for proprietary software; may lead to affordable pricing and increased access [2].

The landscape is characterized by its vast array of high-quality tools spanning numerous domains. From operating systems like Linux to office suites like LibreOffice, and database as PostgreeSQL, FOSS offers solutions that 'rival' proprietary counterparts. Moreover, the collaborative nature of FOSS development ensures continuous improvement and innovation, leading to a diverse ecosystem that serves to various user needs and preferences.

The reasons for FOSS adoption vary from pragmatic to ideological, but they should be based not only on their technical merit, their no-cost feature, or their access to the source code but in terms of technical features, reliability, ease of use, documentation, technical support, customization and extensibility, costs of training, support and maintenance, and management requirements [2].

In complex project development, FOSS plays an indispensable role, offering robust frameworks, libraries, and development tools that facilitate the creation of sophisticated software systems. Whether it is building enterprise-grade applications, managing large-scale infrastructure, or developing cutting-edge technologies, FOSS provides the foundation upon which ambitious projects can succeed. The collaborative nature of FOSS development also fosters knowledge sharing and collective problem-solving, making it well-suited for tackling intricate challenges.

Free and open-source software development cycle allows literally anyone to participate in the process, but having multiple participants means a massive coordination effort (**Figure 1**) [3], being guided by different types of licenses that preserve the authorship of the professionals involved in the development.

Source license models fall into three general categories: free—the program can be freely modified and redistributed; copyleft—the owner gives up intellectual property

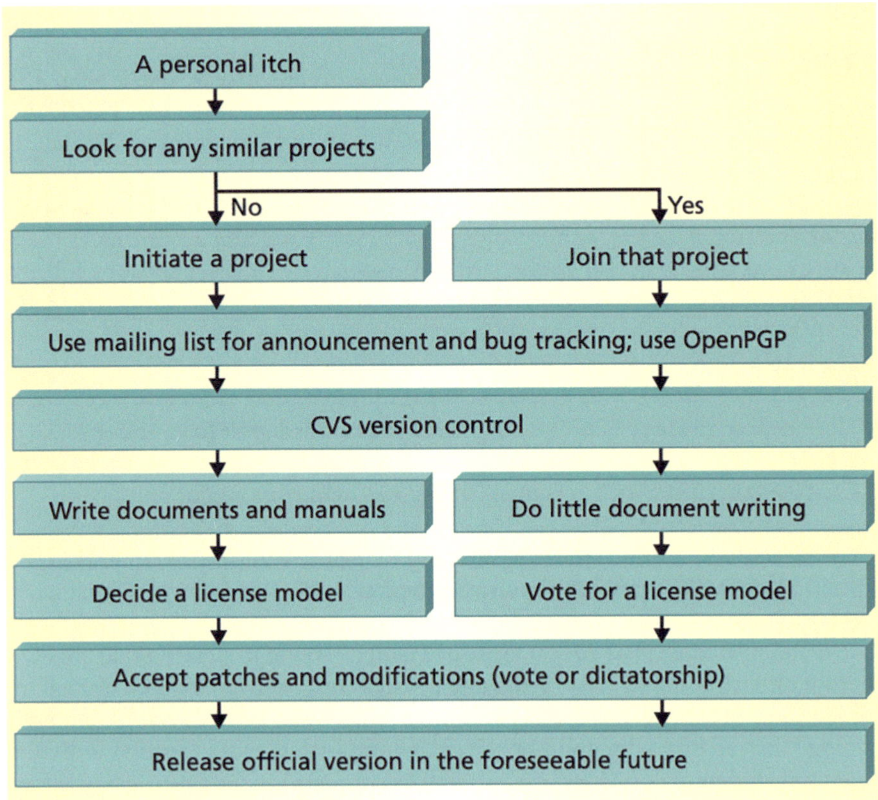

Figure 1.
The general open-source system development cycle [3].

and private licensing; and GPL-compatible—licenses are legally linked to the GPL licensing structure [3], where GPL means General Public License. The license models are just a dimension of the maturity that FOSS has reached and allowed professionals and companies to adopt free and open-source software as a standard for many projects and tasks.

Within these license models, the GPL stands out as one of the most used. The GPL is a copyleft license that allows the modification and redistribution of the software, if the changes are made available under the same license. This promotes collaboration and transparency, as any improvement made to the software can benefit the entire community. In addition, the GPL protects the freedom of users, ensuring that they always have access to the source code and can modify the software to meet their needs. This approach has been fundamental to the growth and maturity of FOSS.

Professionals, ranging from developers and system administrators to data scientists and designers, increasingly rely on FOSS tools and platforms to accomplish their tasks efficiently. Similarly, enterprises recognize the strategic value of embracing FOSS, not only for cost savings but also for fostering innovation, agility, and vendor independence. As FOSS continues to evolve and gain traction, its significance in the modern technological landscape becomes ever more apparent.

2.2 FOSS4G

As well as FOSS, the FOSS4G approach can promote innovation by removing barriers, such as software licensing costs, that tend to surround proprietary computer operating systems and software products, and nowadays, there is at least one mature, sophisticated, free, and open-source software product for every geo-technology area and geospatial application—from data collection in the field, data processing, analysis, modeling and simulations, spatial extensions to database management systems, visualization, and web mapping [1].

In 2006, the Open Source Geospatial Foundation (OSGeo) was created to support and promote the collaborative development of open geospatial technologies and data (http://www.osgeo.org/), and since it happens, there is a growing number of books, online materials, workshops, and commercial services available to learn specific FOSS4G and software stacks [2]. OSGeo was a key milestone to certify the maturity of open geospatial software and the related community [7].

OSGeo hosts an increasing number of software projects, publishes the OSGeo journal, has founded an education and curriculum committee, organizes a yearly international conference (FOSS For Geospatial Conference: FOSS4G; the 2024 Conference will be held in Belém, Brazil), and donates the Sol Katz award to individuals who have demonstrated leadership in the FOS GIS community [6]. FOSS4G conference attracts, each year, more and more participants from research, organizations, and companies.

Projects under the OSGeo umbrella are mature and sustainable, having passed to an incubation process and are certified by OSGeo Incubation Committee, assuring that all projects fulfill the requirements of being geospatial, have an open-source licensing, accept contribution, and accept the code of conduct of OSGeo [1].

The members supporting the projects must belong to several organizations to ensure long-term project viability, while the development leadership (Project Steering Committee) must offer both transparent decision-making and opportunities for new members [1]. **Table 1** shows some of the OSGeo projects.

Under the OSGeo initiatives, QGIS leads as a Desktop GIS software, boasting a user-friendly interface, comprehensive features, and robust community support. It integrates consistently with the Orfeo Toolbox, a potent image processing library for remote sensing applications. This integration enhances QGIS's capabilities, enabling users to conduct advanced geospatial analyses like image classification, object detection, and change detection within the QGIS environment.

Type	Project
Geospatial libraries	GDAL/OGR, PROJ, GEOS, GeoTools, Orfeo ToolBox
Web Mapping GIS	deegree, GeoServer, MapServer, OpenLayers
Spatial Database	PostGIS
Desktop GIS	QGIS, GRASS, gvSIG
Metadata Catalogs, Content Management System	GeoNetwork, GeoNode, OSGeoLive

Table 1.
OSGeo projects by the type of software [1].

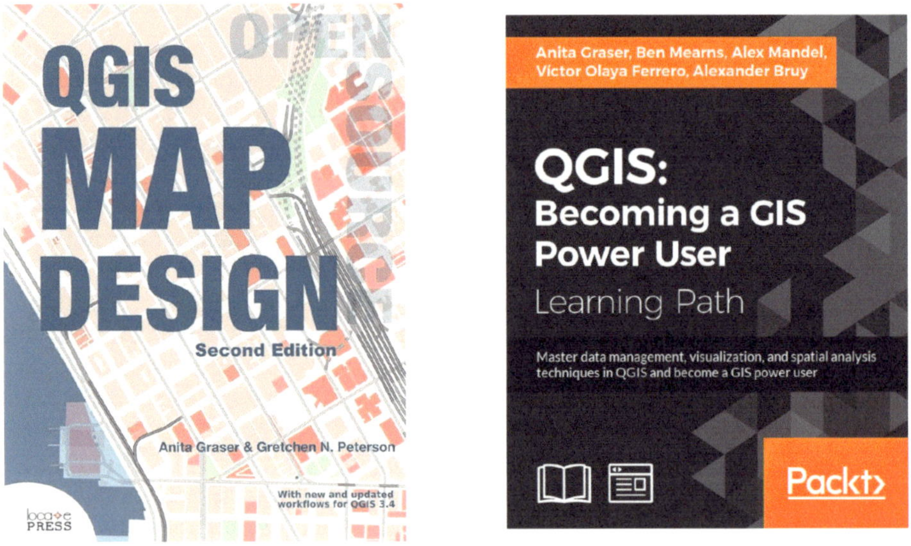

Figure 2.
Two QGIS books examples available for reference and learning.

By utilizing the Orfeo Toolbox's algorithms and QGIS's intuitive interface, users can efficiently process and analyze satellite and aerial imagery. Thus, QGIS serves as a versatile tool for various geospatial tasks.

Beyond its strong analysis capabilities, QGIS provides easy access to basemaps via plugins such as QuickMapService. These basemaps, encompassing well-known sources like OpenStreetMap, Google Maps, and Bing Maps, offer crucial context for visualizing and analyzing spatial data. The QuickMapService plugin streamlines the addition of basemaps to QGIS projects, enabling users to swiftly search and choose from diverse basemap providers. Serving both reference and presentation needs, basemaps amplify the visual and informational quality of QGIS projects, equipping users to create attractive maps and effectively present their results (**Figure 2**).

At the time of writing, there are over 20 books available in digital or print format on QGIS. These include titles like "Exploring the Earth with QGIS: A Guide to Utilizing Satellite Imagery to its Fullest Potential," "QGIS: Advancing to GIS Power User Status," "Creating a Thematic Map in 60 Minutes: Examples Using QGIS," "QGIS Applications in Agriculture and Forestry (Earth System - Environmental Sciences: QGIS in Remote Sensing)," and "QGIS Map Design."

3. Main software for developing geospatial projects

3.1 QGIS

The GIS software, always being defined as Desktop GIS, is the software that is not executed on a server and remotely accessed or controlled from or by a different computer [6] that runs on a personal computer to perform geospatial tasks and analysis.

Quantum GIS or QGIS is an opensource GIS written in C++ available under the conditions of the GPL license that supports a large number of vector and raster formats and provides a plugin mechanism which can be used to add support for a new data source or to extend the functionality of the main program in a modular way (**Figure 3**) [8].

Started in 2002, the QGIS software is currently the most used open-source Desktop GIS application in the world. QGIS project evolved from the idea emerged to use QGIS as a simple Graphical User Interface (GUI) for GRASS (Geographic Resources Analysis Support System) itself [6].

GRASS GIS project first sprouted 30 years ago when the first libraries as an alternative to ESRI's ARCINFO software, developed by the US Army Construction Engineering Research Lab (CERL), with the first release version 1.0 available in 1985 [6]. GRASS was developed to perform geospatial analysis and scientific data visualization, professional cartography, modeling, and simulation.

GIS software, being an open-source software, is particularly appreciated for its availability in the studio and for professional use [9].

QGIS is a powerful tool for data manipulation, georeferencing (**Figure 4**), and vectorization. It supports a wide range of formats, including GeoPackage, a modern format for geospatial data. GeoPackage allows users to store multiple vector and raster geospatial datasets in a single SQLite database file, making data management more efficient.

GeoPackage (*.gpkg) is an open, standard-based, application and platform independent, developed by OGC (Open Geospatial Consortium), which was carefully designed to facilitate widespread adoption and use of a single simple file format by both commercial and open-source software applications on enterprise production platforms as well as mobile hand-held devices [10].

There are numerous resources available for learning QGIS (**Figure 2**), ranging from official documentation and textbooks to online tutorials and video courses. These materials cover a wide range of topics and are designed to help users of all skill levels get the most out of the software.

QGIS's functionality can be significantly extended through plugins. These plugins, developed by the QGIS community, provide additional tools for analysis, data visualization, and a great increase in spatial analysis or processing. They can be easily installed through the QGIS Plugin Manager, allowing users to customize their QGIS environment to suit their specific needs.

QGIS includes a graphical model builder that allows users to automate their workflows (**Figures 5** and **6**). This feature lets users graphically design complex models using a simple drag-and-drop interface. The logic behind these models is straightforward: inputs are processed through algorithms to produce outputs. This powerful tool can save time and reduce the potential for error in repetitive tasks.

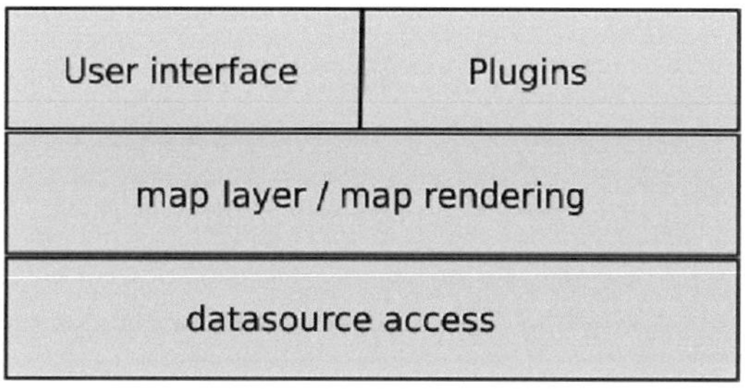

Figure 3.
QGIS code organization [8].

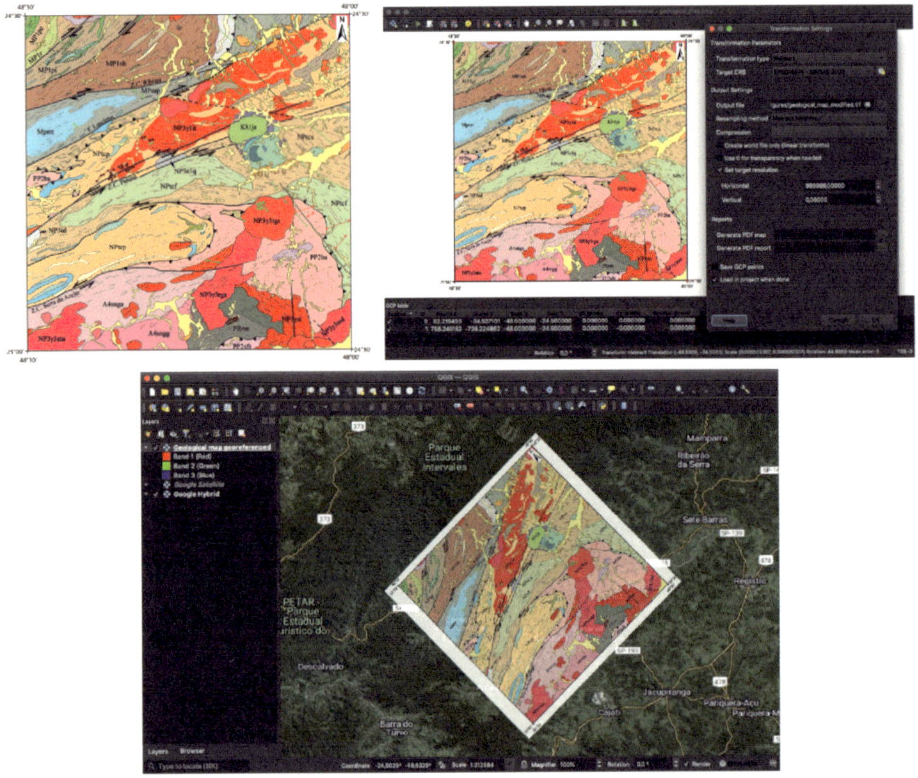

Figure 4.
Georeferencing of a geological map (top left figure), using two points and Helmert transformation type (top right figure) in QGIS georeference tool, and the georeferenced image in QGIS with a Google basemap imagery.

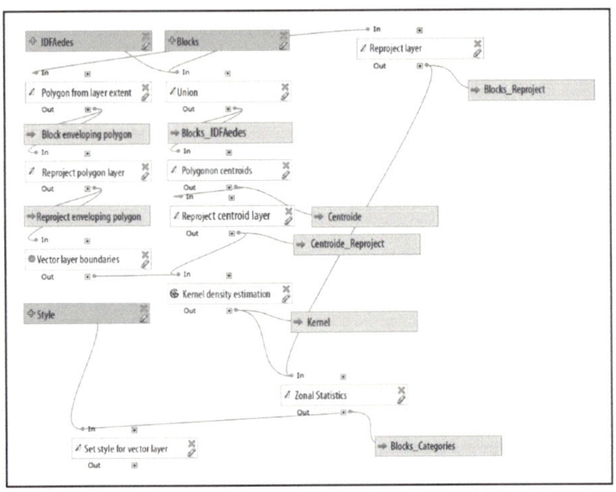

Figure 5.
Example of a QGIS model builder workflow for automatic stratification of priority areas for dengue control [11].

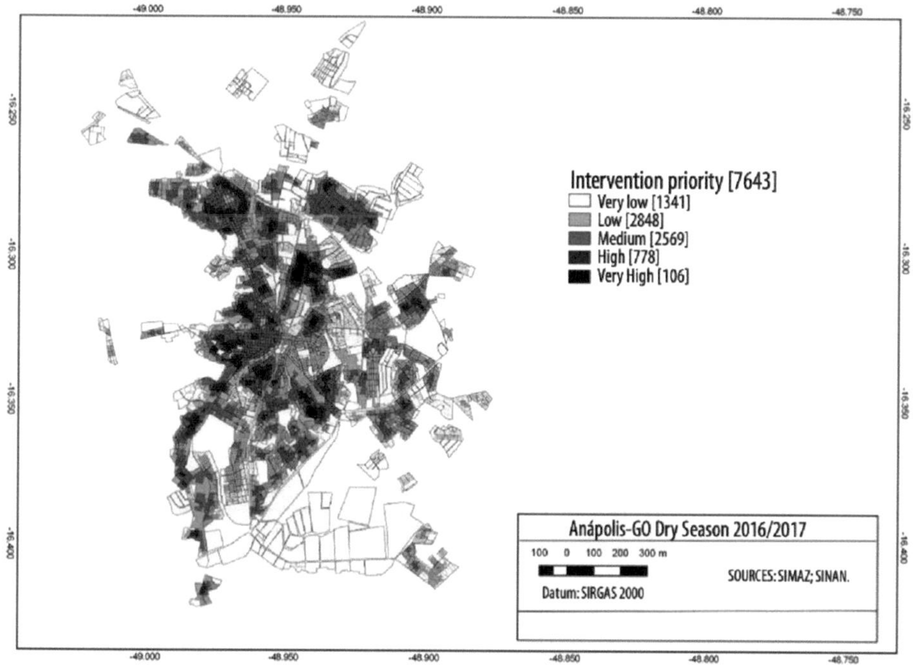

Figure 6.
Result of the automatic selection of priority areas for intervention and planning actions against the Aedes aegypti vector, Anápolis, Goiás, Brazil [11].

3.2 Orfeo Toolbox

Orfeo ToolBox (OTB) is a freely available project dedicated to the processing of satellite images. It is a product of the collective efforts of the open-source geospatial community and is capable of handling high-resolution optical, multispectral, hyperspectral, and radar images, even those of terabyte size. It offers a broad range of applications such as ortho-rectification, pan-sharpening (**Figure 7**), supervised and unsupervised classification, feature extraction, and SAR processing [12]. This application is accessible through the QGIS plugin repository, requiring specific software installations. The installation guide can be found in [13].

With more than 10 years of development and 28 major releases, OTB has evolved from a very-high-resolution specific toolbox driven by the CNES (French Space Agency) Pleiades mission to a multisensor, multipurpose remote sensing toolbox [12]. The goal of OTB is to enable the user to process remote sensing images from different sources (satellite, image provider) with different levels of preprocessing (ortho-rectification and radiometric corrections) [14].

OTB algorithms cover many features needed to process remote sensing images from basic pre-processing to high performance analysis. OTB's mission statement is to provide a free software end-to-end solution for the Earth observation image information extraction pipeline based on a generic, high performance C++ library [12].

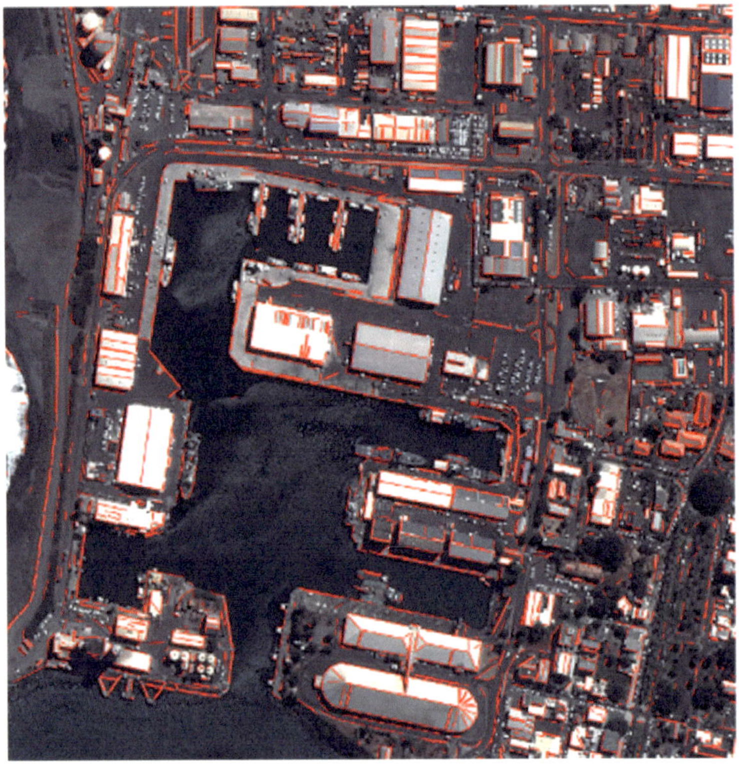

Figure 7.
Results of the Orfeo line segment detector (LSD) applied on a Pleiades pan-sharpened image [12].

The cross platform, ready-to-use OTB package ships with more than 90 applications that is now part of several large, operational projects while still being a tool of choice for the day-to-day image processing in labs and a great environment to write new algorithms and methods. Scalability, versatility, and openness are the key assets of OTB. The software is constantly evolving to enhance performances, to offer more state-of-the-art algorithms and to provide the best interoperability between them [12].

OTB allows has an infinitude of classes that support most of the usual operations on remote sensing images, like image access (optimized read/write access for most of remote sensing raster and vector formats, meta-data access, and visualization), geometric modeling (sensor models, DEM access, cartographic projections, image registration, and disparity map estimation), filtering (blurring, denoising, and enhancement), feature extraction (radiometric indices, textures, interest points, alignments, and lines), image segmentation (region growing, watershed, and level sets), and object oriented analysis, object extraction (road network extraction and template-based detection), classification (K-means, SVM, and Markov random fields), change detection, and on-line data access [14].

OTB can be accessed through the QGIS plugin area, and the user can access the functionalities of the Processing Toolbox to leverage the comprehensive suite of image processing operations provided by the Orfeo Toolbox (**Figure 8**). This integration facilitates a streamlined workflow, enabling users to perform sophisticated remote sensing tasks. The synergy between OTB and QGIS empowers users to efficiently process high-resolution satellite imagery, unlocking the potential for advanced geospatial projects and research.

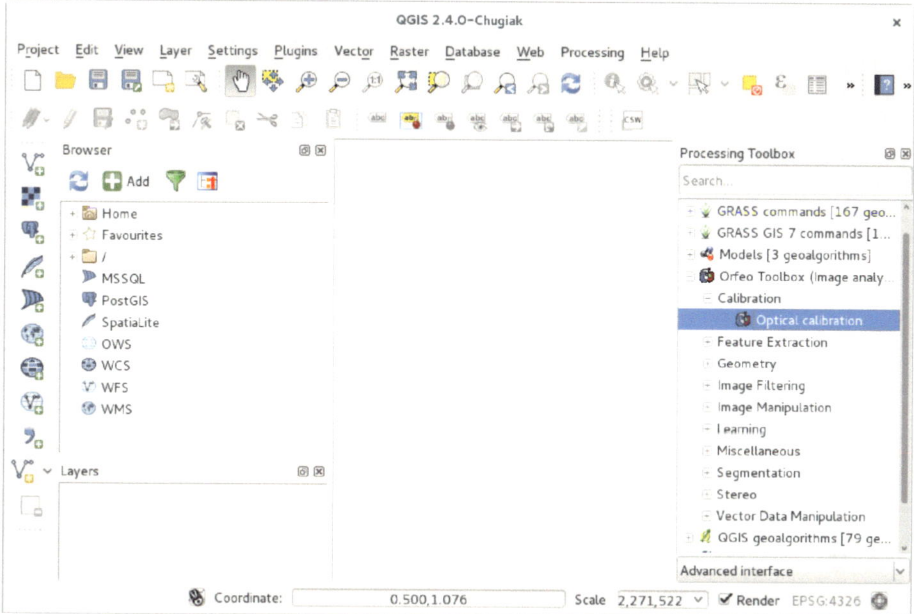

Figure 8.
OTB tools available in the QGIS processing toolbox [13].

3.3 QGIS plugins and basemap options

3.3.1 Plugins tools

Moreover, thanks to its API, it enables expanding functionality through the development of Python plugins, where in recent years, many QGIS plugins for image processing have been developed [15].

QGIS follows a modular approach by providing a plugin mechanism to extend the functionality of the main program, where the plugins can add and remove buttons and menus, and may react to mouse events on the map window [8], increasing the interface functionalities of QGIS.

Starting from May 2024, it is observed that approximately 1249 plugins options are available for installation, as shown in **Figure 9**. These plugins will become accessible through the graphical interface of QGIS. The strategic implementation of plugins significantly enhances the Desktop GIS environment, offering a multitude of specialized applications developed by extensive communities of developers.

3.3.2 QuickMapService

The QuickMapService plugin for QGIS provides access to various satellite imagery services, such as Bing Aerial, Google Satellite, ESRI Images, Waze, and OpenStreetMap (OSM). These services are conveniently accessible within the QGIS interface, as illustrated in **Figure 10**.

QuickMapService demonstrates the integral link between geospatial basemap imagery and user data. It highlights the value of comprehensive geospatial databases, particularly OpenStreetMap (OSM). Since 2004, OSM has been a pivotal open-source platform, providing reliable geospatial data that supports not only QGIS applications

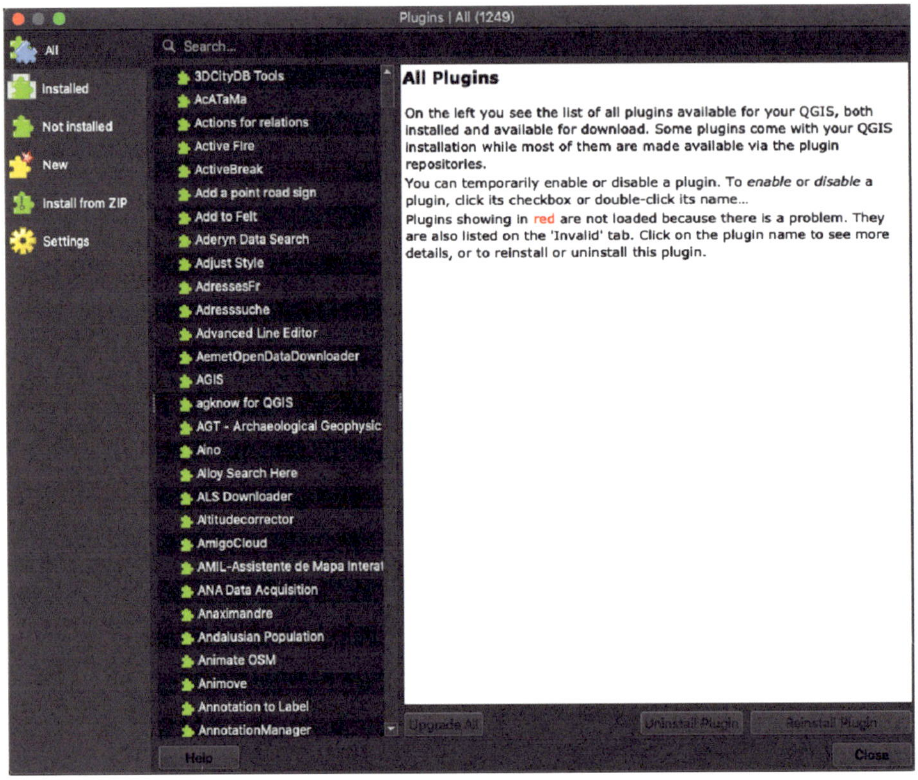

Figure 9.
QGIS plugins available for download and installation.

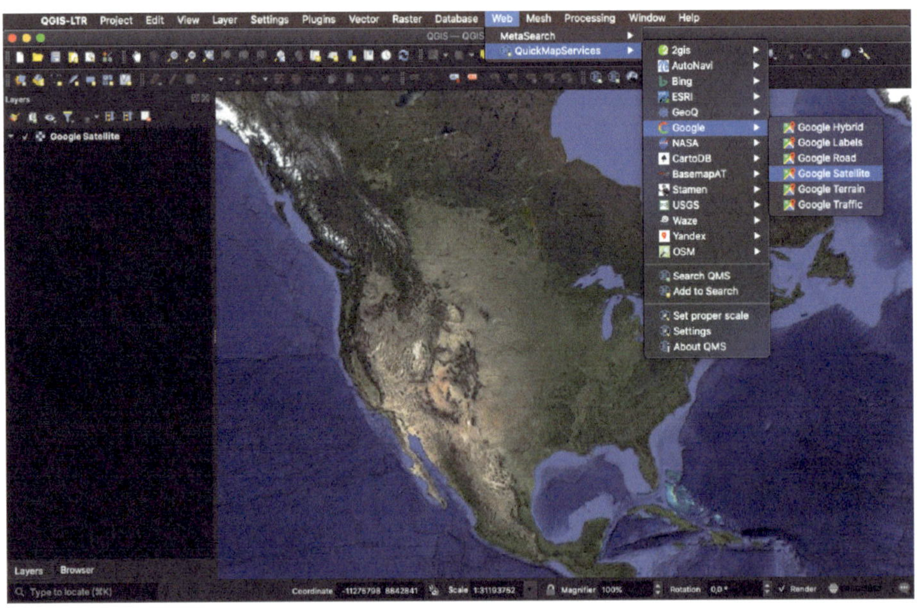

Figure 10.
QuickMapService interface on QGIS showing all the basemap services available.

DOI: http://dx.doi.org/10.5772/intechopen.1005682

but also a wide array of real-world and scholarly activities. The synergy between data and tools like QuickMapService enriches the geospatial analysis landscape.

OSM is an open-access, crowdsourced geospatial database, having a global extent and including any real-world feature that is verifiable on the ground and is currently used in a multitude of real-world and scientific applications as well as deeply studied by the academic community [7]. In addition to providing OSM as a basemap (**Figure 11**), QGIS features a plugin that enables users to download OSM data in vector format. This includes detailed elements such as streets, blocks, buildings, and more, allowing for a more comprehensive and customizable geospatial analysis experience.

Some OSM data are donated by authorities, but mostly, it is collected by amateur enthusiasts with variable geoscientific knowledge and skills, and as a result, the data may vary in quality (completeness, accuracy, precision, etc.,) over different regions [1].

3.4 Increasing analysis with python and open-source libraries

Python is an unparalleled open-source prominence scripting and programming language in geospatial world!!!

Python is the leading scripting and programming language for both proprietary and open-source geospatial software, and Python-based open-source programming has become a key tool in remote sensing-based science [16] and in other geospatial analyses.

The ascendancy of Python in geospatial and remote sensing-based science is a testament to its robust capabilities and quality. With Python at their disposal, researchers and professionals can harness the power of libraries such as OGR/GDAL, PROJ, GeoPandas, Fiona, and Rasterio, to manipulate spatial data, extract meaningful insights, and contribute to the collective understanding of our planet.

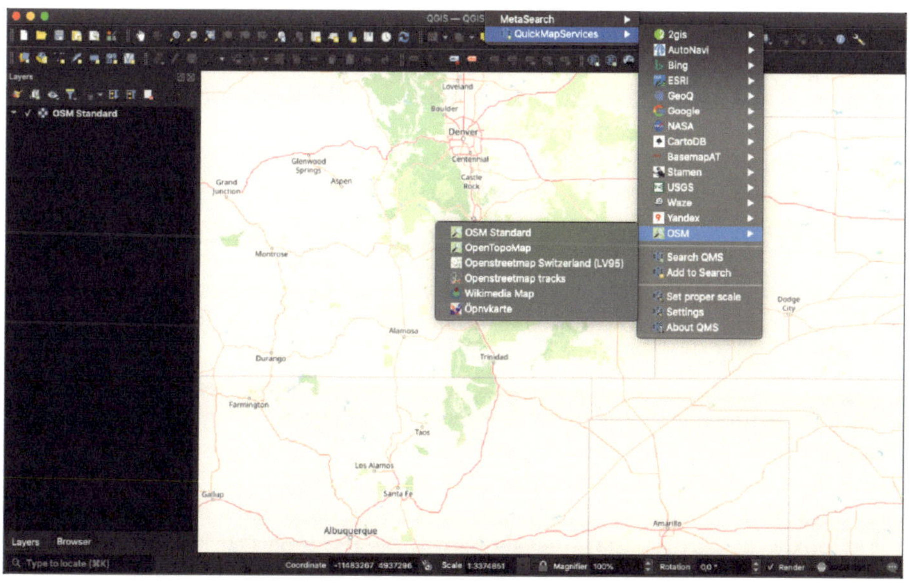

Figure 11.
OSM services available through QuickMapService.

Python's design philosophy prioritizes readability and accessibility, which democratizes programming and empowers a broader audience to engage with geospatial data. This accessibility is crucial in educational settings, where Python serves as a gateway for students and enthusiasts to explore the realm of geospatial analysis.

The ecosystem of Python libraries and packages is a treasure trove for geospatial professionals. Libraries such as NumPy, Pandas, and Matplotlib integrate seamlessly with geospatial libraries, providing a comprehensive toolkit for data processing and visualization. The ability to share findings and insights openly aligns with the ethos of the geospatial community, fostering a culture of knowledge exchange and innovation, that can be used in private and educational projects, increasing the capacity analysis, saving time, and improving performance.

4. Case studies

4.1 Spatial analysis

Spatial analysis in GIS is a powerful process that allows the examination and understanding of geographic data to reveal spatial relationships, patterns, and trends. It extends beyond mere mapping to enable the exploration of spatial characteristics and the interactions between different locations. Employing a variety of spatial operators and analytical tools, this analysis helps in solving complex location-oriented problems and supports decision-making. It involves combining data from multiple sources to derive new insights, calculating the suitability of locations for specific activities or phenomena, and even detecting changes over time through image analysis.

A location-specific phenomenon that can be covered by spatial analysis is the occurrence of landslides. This type of analysis is critical in identifying areas at risk by examining factors such as soil composition, slope gradient, land use, and historical landslide locations. By integrating these variables, GIS can predict potential landslide areas and help in the development of early warning systems.

Landslide susceptibility assessment and mapping are an essential tool in landslide risk management, supporting authorities, practitioners, and decision makers in the more appropriate and sustainable land planning and risk mitigation strategy, and it can be developed by open-source GIS tools.

A map algebra approach for an area in Italy was used in QGIS to create a landslide susceptibility map with the interaction of several predisposing factors like lithology, incoherent soils, heterogeneous materials of anthropic origins, slope aspect and steepness, and land use, properly reclassified (**Figure 12**).

The QGIS raster calculator, as shown in **Figure 13**, enables the execution of equations that represent landslide conditions. In this process, variables are replaced by corresponding maps, culminating in the creation of Landslide Susceptibility Maps (LSM).

The LSM delineates spatial regions with varying degrees of landslide susceptibility, ranging from very low to very high (**Figure 14**). The LSM illustrates the capability of these tools to characterize and quantify phenomena that have significant impacts on our lives [17].

Previously, access to spatial analysis algorithms and tools was limited to those who could afford commercial licenses for proprietary software. Now, these powerful resources are accessible for both professionals and researchers, empowering them to tackle challenges ranging from the simplest tasks to the most complex issues.

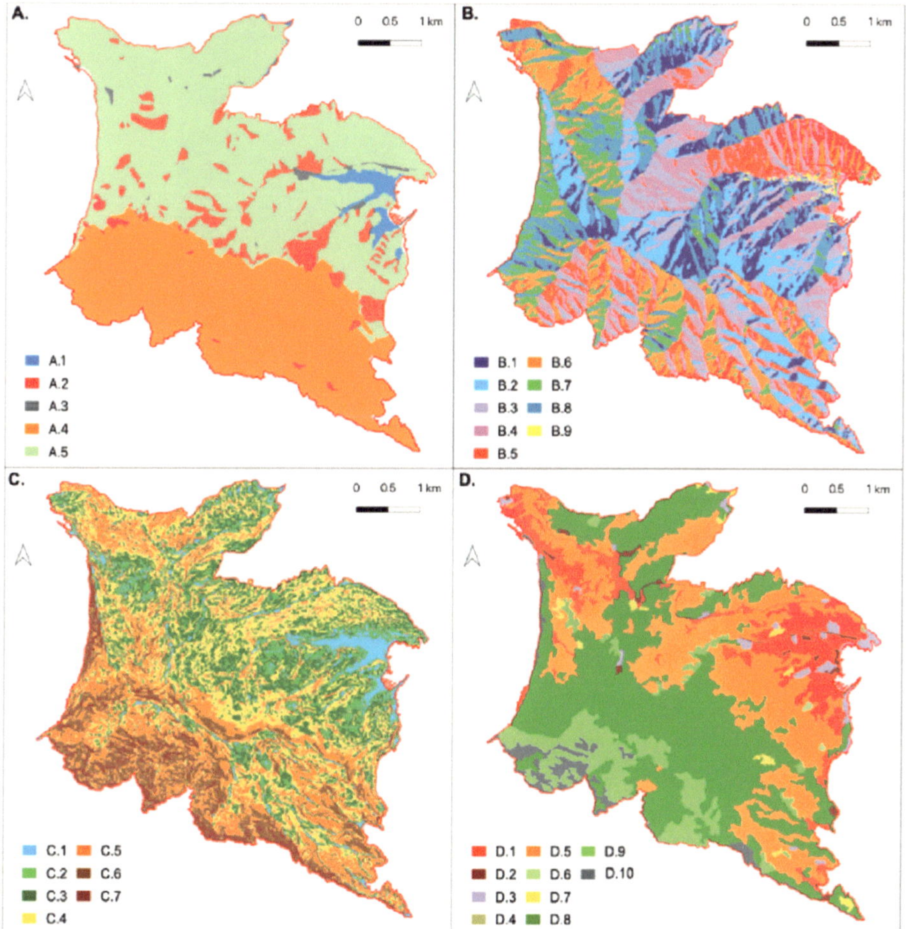

Figure 12.
Shallow landslides predisposing factors with spatial specific classes [17].

4.2 Remote sensing

Remote sensing is a non-intrusive measurement approach used to gather data about objects or areas from a distance, primarily through satellite imagery, aerial photography from planes, and data capture using unmanned aerial vehicles (UAVs). This technology has revolutionized the way we collect and analyze environmental data, offering insights that are critical for various applications such as land surveying, urban planning, and disaster management. The ability to obtain high-resolution images and detailed vector data enhances our understanding of the Earth's surface, making remote sensing an indispensable tool in modern geospatial science.

4.2.1 Image classification

Image classification, a fundamental task in computer vision, involves assigning predefined labels or categories to images based on their visual content. There are many artificial intelligence (AI) models used for satellite imagery classification,

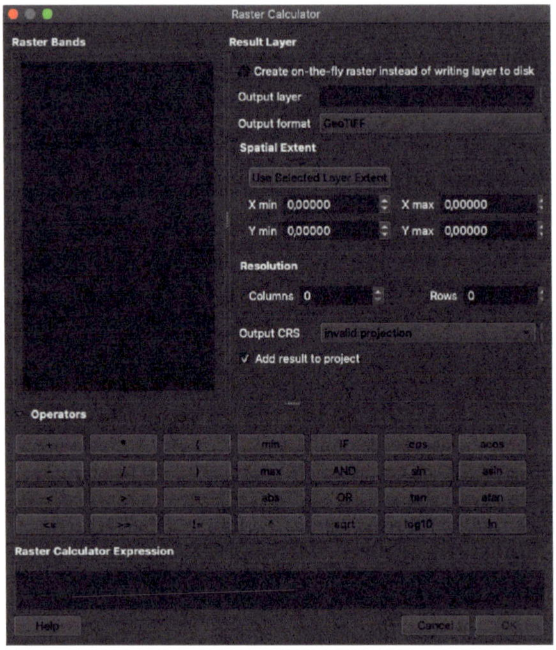

Figure 13.
Raster calculator environment in QGIS, where each variable of a equation can be a raster map.

including machine learning models, for instance Random Forest, XGBoost, SVM, and deep learning models, such as Neural Networks.

For example, Souza et al. [18] used Random Forest models to classify Landsat imagery providing three decades of Brazilian Land Use Land Cover (LULC) (**Figure 15**), while Ferreira et al. [19] used deep learning models to classify data cubes of imagery covering the Brazilian territory (**Figure 16**).

4.2.2 Object detection

Artificial intelligence (AI) revolutionizes the field of remote sensing through cutting-edge object detection models, including YOLO, Faster R-CNN, and Single Shot Multibox Detector (SSD). These sophisticated models utilize deep learning algorithms to sift through extensive datasets, enabling the automated extraction of critical environmental information from satellite and aerial imagery. AI algorithms, fortified by techniques such as transfer learning, data augmentation, and attention mechanisms, adeptly navigate the complexities of remote sensing. Challenges like image resolution variability, scale differences, and the scarcity of annotated data are deftly managed. The synergy of multi-scale feature fusion and semantic segmentation within these AI-driven object detection systems culminate in unparalleled precision, identifying a wide array of objects from vehicles and edifices to flora and geological formations with remarkable accuracy.

Furthermore, the amalgamation of Python libraries such as Keras, TensorFlow, OpenCV, and PyTorch with avant-garde object detection frameworks equips researchers and developers with the tools to unleash the full potential of artificial intelligence. This empowerment extends beyond remote sensing, facilitating object detection endeavors across diverse applications.

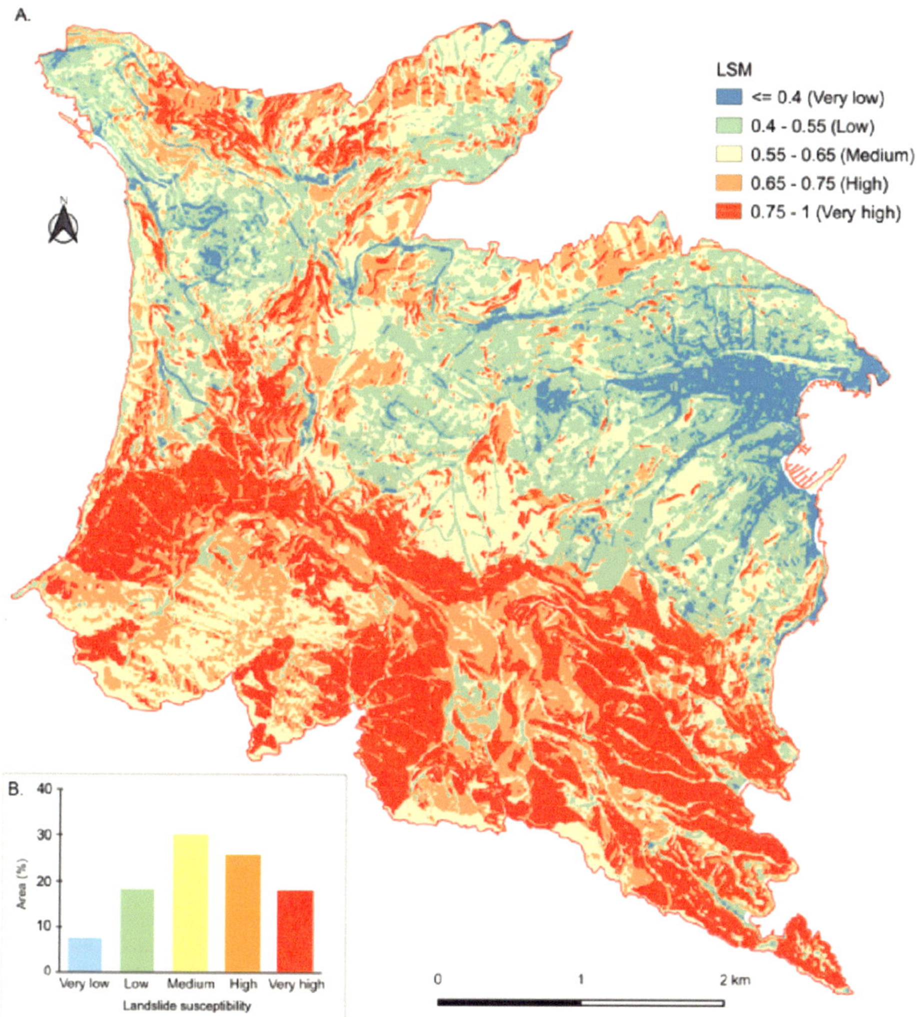

Figure 14.
(A) Landslide susceptibility map and (B) histogram of the percentage of areas included in the five susceptibility classes [17].

A transformative approach to ship detection has been realized through the adoption of the Faster R-CNN architecture [20]. This method marks a significant advancement in identifying small and clustered ships, as demonstrated in **Figure 17**. The key to this enhancement is the strategic refinement of the VGG16 CNN framework, with the incorporation of multiresolution convolutional features, and implementing ROI pooling across a larger feature map, the detection process has been finely tuned for optimal performance.

In the realm of building footprint extraction, the Google Open Buildings Dataset (OBD), which is open-source, stands out for its global reach and deep learning-driven methodology. The OBD employs state-of-the-art convolutional neural networks (CNNs) and deep learning strategies to autonomously extract building footprints and metadata from satellite imagery [21]. The dataset, depicted in **Figure 18**, is a testament to the efficacy of models trained on extensive datasets, showcasing their ability to precisely identify and outline building structures from satellite images.

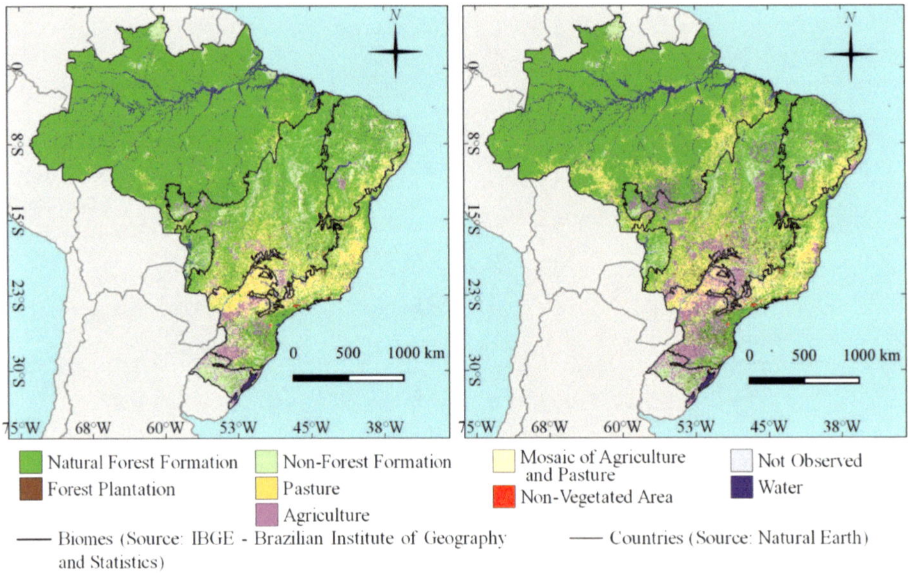

Figure 15.
Example of random Forest for LULC: (a) LULC 1987, and (b) LULC 2017 [18].

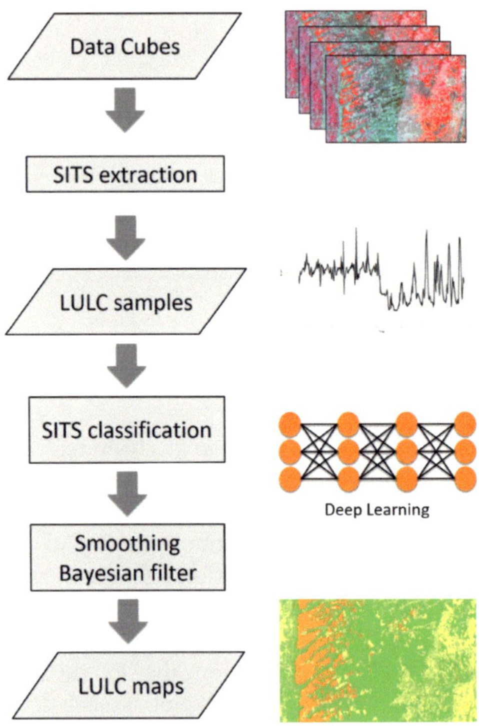

Figure 16.
Example deep learning for LULC [19].

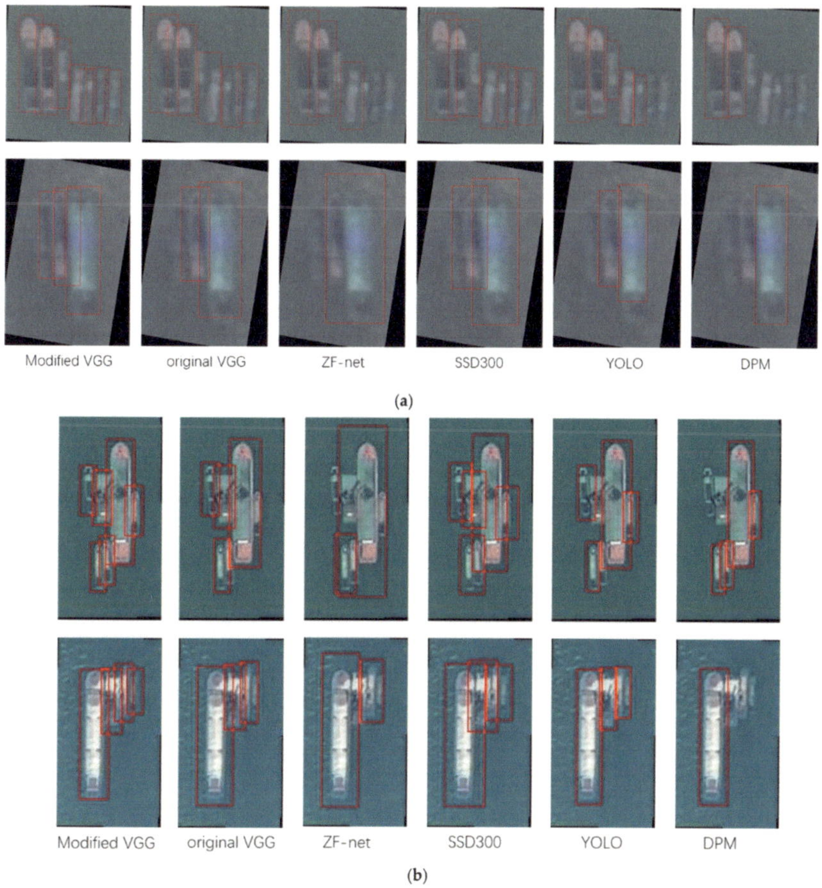

Figure 17.
Example of ships detection [20].

Following the capacity of OBD, a QGIS plugin was developed. Deepness is a plugin (**Figure 19**) that allows the easy employment of neural network models on any raster layer representing a matrix of values or image data (**Figure 20**) and connects deep learning expertise with remote sensing applications' needs and is targeted and designed to be operated by three different groups of people: casual users, model creators, and developers [15].

The primary function of the Deepness is to execute inference using custom neural network models on any raster layer or combination of raster layers in QGIS. It offers three different modes for processing areas:

- Processing the entire raster layer file;
- Processing only the area confined within a polygon, which allows for the specification of subregions;
- Processing only the portion currently visible on the screen, which facilitates quick usage while viewing large orthophotos, including massive maps.

Figure 18.
Example of google open buildings dataset [21].

It also supports three prevalent types of deep learning models: (a) segmentation, (b) regression, and (c) detection. Segmentation calculates the likelihood of each pixel belonging to a predefined class, based on its local and global context. In regression, the network predicts a single continuous value for the regressed variable for each pixel. Detection identifies and classifies objects in an image via their bounding boxes (**Figure 21**).

4.2.3 Monitoring water quality using remote sensing and open geospatial tools

Water quality is a fundamental component of environmental health, impacting not only ecosystems but also human health and socio-economic development. Monitoring water quality is crucial for ensuring the availability of clean and safe water for various uses, including drinking, agriculture, industry, and recreation.

Traditional methods of water quality monitoring often involve time-consuming and labor-intensive field sampling. However, remote sensing technology offers a powerful alternative, allowing for the efficient and comprehensive assessment of water quality over large spatial scales. By utilizing satellites and other remote sensing platforms, scientists can collect valuable data on parameters such as turbidity, chlorophyll concentration, and water temperature, providing insights into the health of aquatic ecosystems and aiding in the management of water resources.

Several open resources of remote sensing tools facilitate water quality monitoring on a global scale. One prominent example is NASA's Earth Observing System (EOS), which includes satellites such as Landsat and Aqua that provide multispectral imagery suitable for assessing water quality parameters.

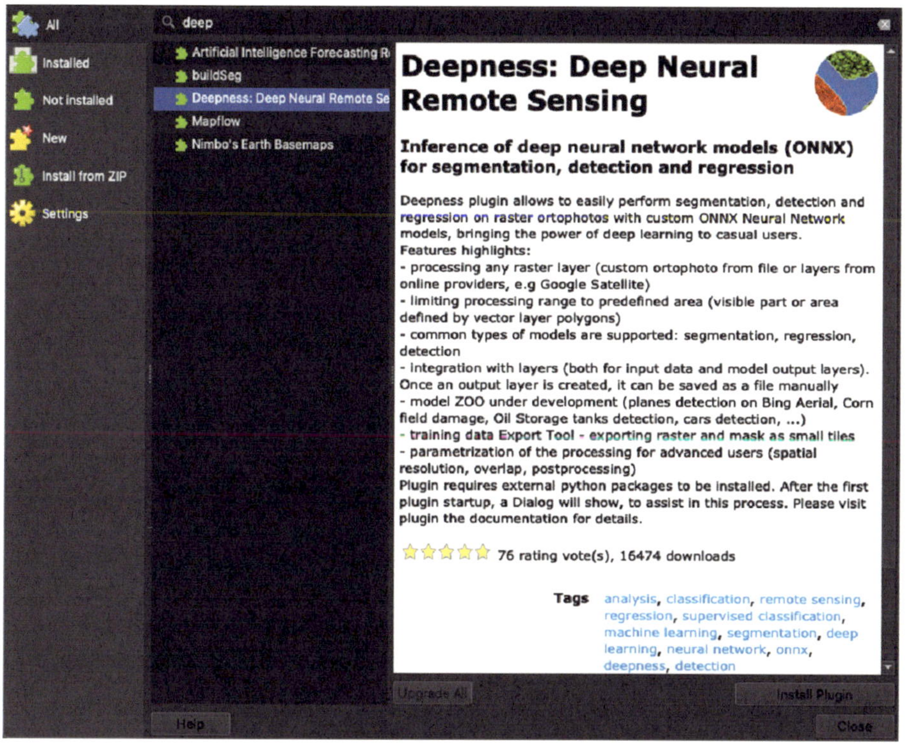

Figure 19.
Deepness plugin is available in QGIS plugin manager [21].

The European Space Agency's (ESA) Sentinel satellites also offer freely accessible data for monitoring water quality, with sensors capable of detecting various water quality indicators. Additionally, initiatives like Google Earth Engine provide a platform for accessing and analyzing vast repositories of remote sensing data, enabling researchers and policymakers to monitor water quality trends and make informed decisions regarding water resource management [22].

These open resources democratize access to remote sensing data, empowering stakeholders worldwide to contribute to the preservation and sustainable management of water resources. Moreover, libraries like geemap, built on platforms like Google Earth Engine, enhance accessibility and usability by providing user-friendly interfaces and tools for data visualization, analysis, and sharing. Geemap, as detailed in [23], facilitates the integration of Google Earth Engine with the Python programming language, opening opportunities for a broader community of users to leverage remote sensing data for water quality monitoring and other environmental applications.

One critical step in remote sensing is atmospheric correction, which accounts for the effects of the Earth's atmosphere on remote sensing measurements. Atmospheric correction algorithms aim to remove atmospheric interference such as scattering and absorption, thereby improving the accuracy of water quality parameters derived from remote sensing data. Some atmospheric correction methods are publicly available (under AGPL-3.0 license) and can be accessed through by https://github.com/MarcYin/SIAC [24].

While remote sensing technologies provide valuable images for water quality assessment, further processing and analysis are essential to derive accurate and

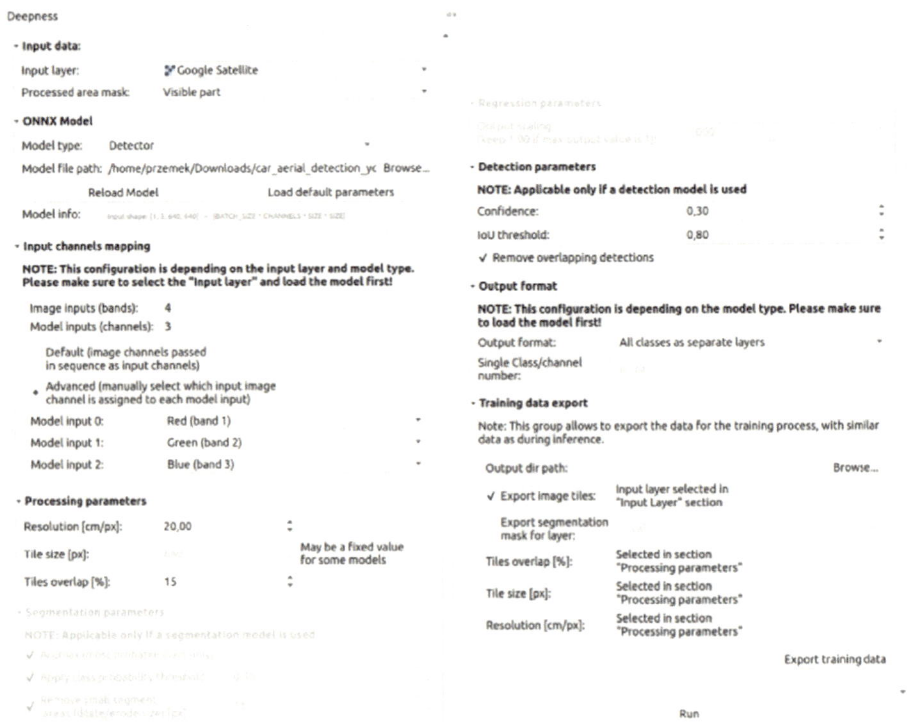

Figure 20.
View of the main deepness plugin UI, allowing to select all essential parameters and run the processing [15].

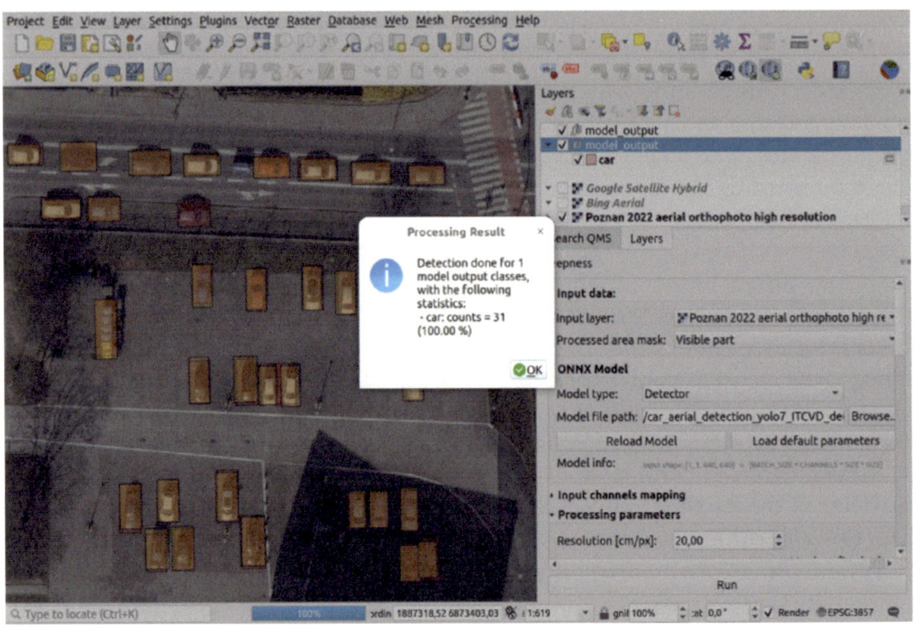

Figure 21.
Example of cars detection at Poznan university technology [15].

meaningful information. Data analysis open-source libraries like pandas [25], NumPy [26], and matplotlib [27] facilitate data manipulation, visualization, and statistical analysis, enabling researchers to explore relationships between remote sensing variables and water quality parameters. Others such as scikit-learn [28], TensorFlow [29], and PyTorch [30] provide robust frameworks for developing machine learning models to predict water quality indicators from remote sensing imagery. These libraries offer a wide range of algorithms, including regression, classification, and clustering techniques, which can be trained on remote sensing data along with ground truth measurements to generate predictive models.

A method to predict water quality at an urban reservoir located in Belo Horizonte, Minas Gerais, Brazil, was developed using Sentinel-2 and Landsat-8 images corrected with SIAC [31]. The reflectance values were combined with in situ water quality measures of Total Suspended Matter (TSM), Chlorophyll-a (Chla), Secchi Disk Depth (SDD) and Turbidity. The methods were organized and integrated using the following Python libraries:

- Data analysis: pandas, numpy, geopandas
- Model: scikit-learn
- Remote Sensing: Google Earth Engine Python API, xarray, rasterio
- Visualization: matplotlib.

Figure 22 shows the modeled water quality parameters for the reservoir and how Python allowed to organize, analyze, predict, and visualize it.

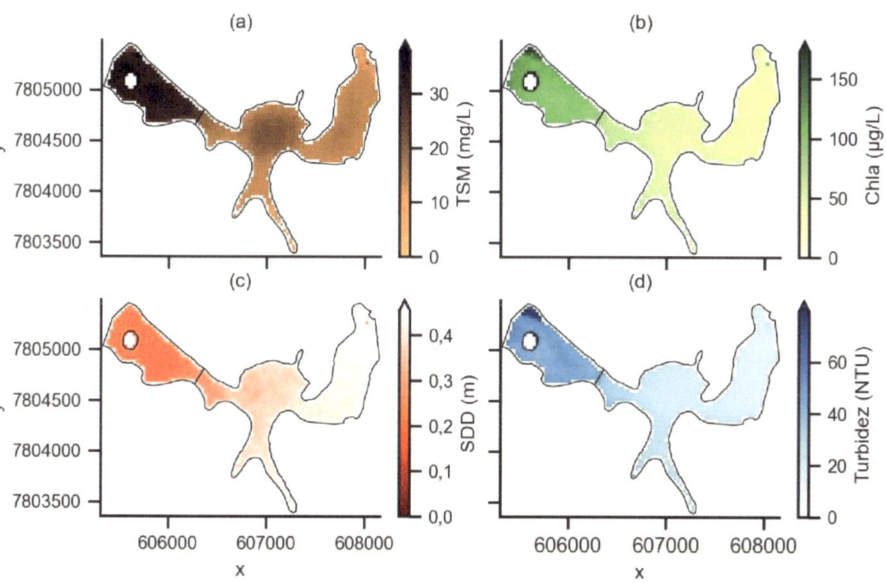

Figure 22.
Median modeled values aggregated over time (a) TSM, (b) Chla, (c) SDD, and (d) turbidity [31].

5. Conclusions

Over the past few decades, initiatives like FOSS (Free and Open Source Software) and FOSS4G (Free and Open Source Software for Geospatial) have gained significant momentum. These initiatives offer high-quality, mature software options without the onus of licensing costs. They empower the development of geospatial projects, allowing professionals to tackle complex tasks. What began as an "amateur" endeavor, fueled by developer and community contributions, has now evolved into institutional support that professionalizes actions and leverages geospatial tools.

Regarding FOSS4G, the Open Source Geospatial Foundation (OSGeo) oversees various software projects. Additionally, OSGeo promotes learning materials and hosts an annual global conference. At this conference, professionals, researchers, and developers come together to present technical work and case studies. This collaborative environment contributes to the software's maturity and expands its user base.

The highlighted tools in this context enable geospatial professionals to undertake projects of varying complexity and scope, as demonstrated in the Case Studies section. By offering a viable alternative to proprietary and expensive tools, these open-source solutions empower both research and market-oriented projects.

The primary goal of this chapter is to introduce readers to the concepts of FOSS and FOSS4G, presenting a fresh perspective and diverse options for geospatial project development. Accessible tools and educational resources further facilitate the adoption of new software, reducing the learning curve for users.

Acknowledgements

I extend my heartfelt gratitude to all the dedicated individuals who, through unwavering belief and substantial effort, have elevated FOSS (Free and Open Source Software) and FOSS4G (Free and Open Source Software for Geospatial) tools to their current state. Their invaluable contributions have empowered the authors of this chapter, as well as a vast user community, to undertake numerous research and market projects, leveraging the wealth of resources available in the free geospatial domain.

Author details

Leonardo Santana de Oliveira Dias*†, Alexandre Flávio Assunção† and Felipe Augusto Nascimento de Jesus†
Tetra Tech South America, Belo Horizonte, Brazil

*Address all correspondence to: leonardo.santana@tetratech.com

† These authors contributed equally.

References

[1] Coetzee S, Ivánová I, Mitasova H, Brovelli MA. Open geospatial software and data: A review of the current state and a perspective into the future. ISPRS International Journal of Geo-Information. 2020;**9**(2):90

[2] Moreno-Sanchez R. Free and open source software for geospatial applications (FOSS4G): A mature alternative in the geospatial technologies arena. Transactions in GIS. 2012;**16**(2):81-88

[3] Wu MW, Lin YD. Open source software development: An overview. Computer. 2001;**34**(6):33-38

[4] Gonzalez-Barahona JM. A brief history of free, open source software and its communities. Computer. 2021;**54**(2):75-79

[5] Quinn S. Free and open source GIS in South America: Political inroads and local advocacy. International Journal of Geographical Information Science. 2020;**34**(3):464-483

[6] Steiniger S, Bocher E. An overview on current free and open source desktop GIS developments. International Journal of Geographical Information Science. 2009;**23**(10):1345-1370

[7] Minghini M, Mobasheri A, Rautenbach V, Brovelli MA. Geospatial openness: From software to standards & data. Open Geospatial Data, Software and Standards. 2020;**5**:1-5

[8] Hugentobler M. In: Shekhar S, Xiong H, editors. Quantum GIS. Boston, MA: Springer; 2008. pp. 935-939

[9] De Donatis M, Pappafico GF. Applying a geographic information system and other open-source software to geological mapping and modeling: History and case studies. Geomatics. 2023;**3**(4):465-477

[10] Rashidan MH, Musliman IA. GeoPackage as future ubiquitous GIS data format: A review. Jurnal Teknologi. 2015;**73**(5):47-53

[11] Paiva Júnior EF, Vaz TS, Rosa M, Garcia ILB. Automatic stratification of priority areas for dengue control using the QGIS model builder in multicriteria analysis. Epidemiologia e Serviços de Saúde. 2020;**29**:8

[12] Grizonnet M, Michel J, Poughon V, Inglada J, Savinaud M, Cresson R. Orfeo ToolBox: Open source processing of remote sensing images. Open Geospatial Data, Software and Standards. 2017;**2**(1):15

[13] Orfeo ToolBox. Orfeo Toolbox Documentation. Paris, France: CentreNational d' Etudes Spatiales (CNES); 2024. Available from: https://www.orfeo-toolbox.org/CookBook-6.6.1/QGIS-interface.html [Accessed: May 13, 2024]

[14] Inglada J, Christophe E. The Orfeo Toolbox remote sensing image processing software. In: 2009 IEEE International Geoscience and Remote Sensing Symposium. Vol. 4. Cape Town, South Africa: IEEE Xplore; 2009. pp. IV-733-IV-736

[15] Aszkowski P, Ptak B, Kraft M, Pieczyński D, Drapikowski P. Deepness: Deep neural remote sensing plugin for QGIS. SoftwareX. 2023;**23**:101495

[16] Cortés IM. Open-source software for geospatial analysis. Nature Reviews Earth & Environment. 2023;**4**(3):143-143

[17] Roccati A, Paliaga G, Luino F, Faccini F, Turconi L. GIS-based landslide

susceptibility mapping for land use planning and risk assessment. Land. 2021;**10**(2):28. Available from: https://www.mdpi.com/2073-445X/10/2/162

[18] Souza CM, Shimbo Z, Rosa MR, Parente LL, Alencar A, BFT R, et al. Reconstructing three decades of land use and land cover changes in Brazilian biomes with Landsat archive and earth engine. Remote Sensing. 2020;**12**(17):27. Available from: https://www.mdpi.com/2072-4292/12/17/2735

[19] Ferreira KR, Queiroz GR, Vinhas L, Marujo RFB, Simoes REO, Picoli MCA, et al. Earth observation data cubes for Brazil: Requirements, methodology and products. Remote Sensing. 2020;**12**(24):19. Available from: https://www.mdpi.com/2072-4292/12/24/4033

[20] Zhang S, Wu R, Xu K, Wang J, Sun W. R-CNN-based ship detection from high resolution remote sensing imagery. Remote Sensing. 2019;**11**(6):15. Available from: https:// www.mdpi.com/2072-4292/11/6/631

[21] Sirko W, Kashubin S, Ritter M, Annkah A, Bouchareb YSE, Dauphin Y, et al. Continental-scale building detection from high resolution satellite imagery. New York, USA: Cornell University; 2021. Available from: https://arxiv.org/abs/2107.12283

[22] Gorelick N, Hancher M, Dixon M, Ilyushchenko S, Thau D, Moore R. Google earth engine: Planetary-scale geospatial analysis for everyone. Remote Sensing of Environment. 2017;**202**:18-27

[23] Wu Q. Geemap: A python package for interactive mapping with Google earth engine. Journal of Open Source Software. 2020;**5**(51):2305

[24] Yin F, Lewis PE, Gómez-Dans JL. Bayesian atmospheric correction over land: Sentinel-2/MSI and Landsat 8/OLI. Geoscientific Model Development. 2022;**15**(21):7933-7976

[25] The pandas development team. pandas-dev/pandas: Flexible and powerful data analysis / manipulation library for Python, providing labeled data structures similar to R data.frame objects, statistical functions, and much mores. Zenodo; 2024. Available from: https: //github.com/pandas-dev/pandas [Accessed: January 15, 2024]

[26] Harris CR, Millman KJ, van der Walt SJ, Gommers R, Virtanen P, Cournapeau D, et al. Array programming with NumPy. Nature. 2020;**585**(7825): 357-362

[27] Hunter JD. Matplotlib: A 2D graphics environment. Computing in Science & Engineering. 2007;**9**(3):90-95

[28] Pedregosa F, Varoquaux G, Gramfort A, Michel V, Thirion B, Grisel O, et al. Scikit-learn: Machine learning in python. Journal of Machine Learning Research. 2011;**12**:2825-2830

[29] Abadi M, Agarwal A, Barham P, Brevdo E, Chen Z, Citro C, et al. TensorFlow: Large-Scale Machine Learning on Heterogeneous Systems. 2015. Available from: https://www.tensorflow.org/

[30] Paszke A, Gross S, Massa F, Lerer A, Bradbury J, Chanan G, et al. Pytorch: An imperative style, high-performance deep learning library. Advances in Neural Information Processing Systems. 2019;**32**:12. Available from: https://arxiv.org/pdf/1912.01703

[31] Assunção AF et al. Uso do sensoriamento remoto para avaliação da dinâmica espaço-temporal de parâmetros de qualidade da água de um reservatório urbano [Master thesis]. Belo Horizonte: Universidade Federal de Minas Gerais; 2023

Chapter 2

Machine Learning-Based Active Layer Thickness Estimation over Permafrost Landscapes by Upscaling Airborne Remote Sensing Measurements with Cloud-Computing Geotechnologies

Michael A. Merchant and Lindsay McBlane

Abstract

Earth observation (EO) plays a pivotal role in understanding our planet's rapidly changing environment. Recently, geospatial technologies used to analyse EO data have made remarkable progress, in particular from innovations in Artificial Intelligence (AI) and scalable cloud-computing resources. This chapter presents a brief overview of these developments, with a focus on geospatial "big data." A case study is presented where Google Earth Engine (GEE) was used to upscale airborne active layer thickness (ALT) measurements over an extensive permafrost region. GEE's machine learning (ML) capabilities were leveraged for upscaling measurements to several multi-source satellite EO datasets. Novel Explainable Artificial Intelligence (XAI) techniques were also used for model feature selection and interpretation. The optimized ML model achieved an R^2 of 0.476, although performance varied by ecosystem. This chapter highlights the capabilities of new RS sensors and geospatial technologies for better understanding permafrost environments, which is important in the face of climate change.

Keywords: big data, Google Earth Engine, permafrost, remote sensing, explainable AI

1. Introduction

The introduction of this chapter first examines the portion of soil in cold regions that seasonally freezes and thaws, the active layer. More specifically, the remote sensing (RS) of the active layer variable is overviewed, with a focus on Earth observation (EO) imaging techniques. A case study is then presented, in which the maximum thaw depth of the active layer is estimated for the year 2017 across a large area of Canada's northern permafrost region. The methodology leverages airborne and satellite RS data, and cloud-based machine learning (ML) for scaling the predictions. As such, a short background on cloud computing is presented.

1.1 The active layer of permafrost regions

Permafrost is a key element of the cryosphere, and is defined as "cryotic ground" (soil or rock) that remains at 0°C for two or more consecutive years. Across the northern hemisphere, permafrost underlies an estimated 25% (15 million km^2) of the periglacial landscape [1]. The soil layer above the permafrost which thaws and freezes on an annual basis is referred to as Active Layer Thickness (ALT; [2]; **Figure 1**), and is an area where several essential hydrological, biological, and chemical processes take place [3]. This is because the frozen permafrost layer below is impermeable, resulting in wet conditions, slow decomposition rates, and the accumulation of large organic carbon stores above [4]. ALT, or the maximum thaw depth of this layer, is a critical variable to monitor since it corresponds to the stability and thermal state of the permafrost layer [5]. When permafrost degrades (i.e., thaws) from changing environmental conditions, the active layer extends further into the upper portion of frozen ground. Changes to the thermal state of the ground, and accordingly the evolution of ALT, can result in permafrost layer degradation and significant impacts on terrestrial ecosystem processes [6]. These impacts on the sensitive and vulnerable permafrost layer may have significant consequences, both for local communities living on permafrost [7], as well as the global society [8].

Important variables controlling the seasonal fluctuations and thickness of the active layer include hydrology, soil, vegetation, topography, ecotype, and climatological parameters, as well as anthropogenic activities [2, 9]. In particular, air temperatures and their inter- and intra-annual amplitudes exert a strong control on ALT, especially during the thaw season [10]. As air temperatures become warmer, the seasonal duration and depth of thawing becomes more pronounced. This makes permafrost one of the elements of the Earth system most reactive to climate change [11]. Hence why global warming is of such concern for permafrost loss and deepening of the ALT, as changes to ALT may have profound impacts on polar region hydrology and biogeochemical processes [12]. With the latter, permafrost degradation has the

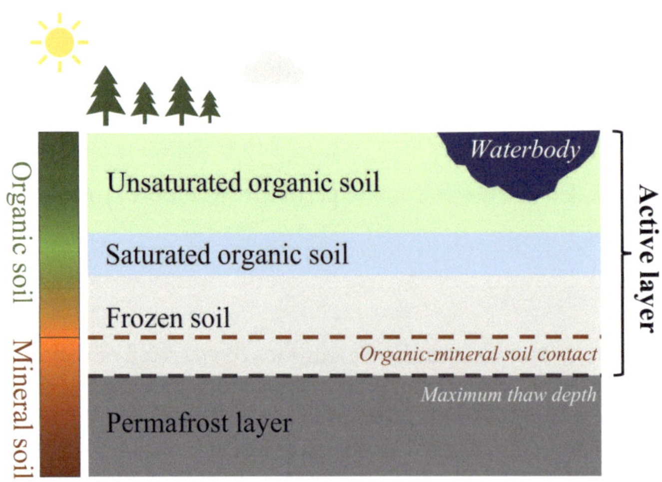

Figure 1.
Conceptual diagram of the active layer in permafrost regions.

potential to release large stores of carbon dioxide (CO^2) and methane (CH^4; [13, 14]). This release may further amplify surface warming at a global scale, resulting in what's referred to as the permafrost carbon feedback (PCF; [15, 16]). Recent research shows that the Arctic has warmed nearly four times the global average [17], thus the PCF and its associated mechanisms require improved understanding, measurements, and modeling in the face of climate change [18].

1.2 Remote sensing of active layer thickness

Generating regional maps of ALT is challenging, since ALT is temporally and spatially variable and reflective of the landscape's heterogeneous hydro-ecological conditions (e.g., snow cover, moisture, vegetation, etc.; [19]). Local scale in situ methods, including ground penetrating radar (GPR), geophysical surveys and conventional human-based sampling (e.g., point-based field measurements from manual frost/thaw probing with metal rods), are proven and effective techniques for determining subsurface permafrost depth [20]. However, these methods are labor intensive, time-consuming, expensive to conduct, and therefore inefficient and not reasonably scalable over large spatial extents, especially across the very remote and inaccessible regions of the Arctic tundra. On the other hand, RS measurements, such as from EO satellite missions, offer a desirable alternative to in situ methods [21]. This is because RS methods are scalable over wide coverages, cost-effective, repeatable, and non-invasive [22].

Several studies have successfully demonstrated the effectiveness of RS (both aerial and satellite) for local-scale ALT estimation and mapping [23], although few have done so over large spatial extents of the northern permafrost region and simultaneously at high resolutions. A number of studies have used optical and/or hyperspectral imaging to map ALT [24–26], whereby retrieval of vegetation coverage is used to investigate ALT conditions. The Normalized Difference Vegetation Index (NDVI) is a well-established technique for understanding ALT with optical RS data. This is because ALT and NDVI are negatively correlated, with greater vegetation coverages (i.e., biomass) resulting in a shallower active layer [27]. Hence, NDVI is often considered as a predictor (i.e., driver) of ALT [28, 29].

Synthetic Aperture Radar (SAR) RS has also become very popular for mapping ALT. This is because the microwaves emitted by a SAR can penetrate vegetation canopies and soil surfaces, and are sensitive to subsurface conditions [30]. These conditions, such as soil organic carbon and moisture, can be used as a proxy for improved inference of ALT [21]. For example, Widhalm et al. [31] demonstrated the interrelations between X-band SAR backscatter and land cover conditions for modeling ALT, with results indicating a better performance than the NDVI variable. Interferometric SAR (InSAR) techniques in particular have proven effective for ALT estimation and monitoring [32]. The InSAR technique relies on both the phase and amplitude information contained in the SAR signal, acquired from a pair (or more) of multi-temporal SAR images [33]. By comparing the timing of the radar return between each image, the distance between the sensor and ground can be quantified. This information is used to infer permafrost conditions, since changes to the vertical distribution of pore water in the active layer is related to InSAR measured subsidence, and thus ALT [34].

1.3 Cloud-based geospatial analysis with Google Earth Engine

The availability of datasets extracted from various RS sources is ever increasing. Rapid technological advancements have resulted in an unprecedented archive of

EO data, which is continuously growing in velocity, volume, and variety [35]. This proliferation of data is referred to as "big data," which is not easily stored or processed using conventional desktop-based resources. Several cloud-based processing platforms have been developed that enable geospatial analysis [36], evidently solving the challenge of big data processing. Most prominent and widely used is Google Earth Engine (GEE), a cloud-based platform that facilitates the analysis of RS data over large spatial extents, by leveraging Google's infrastructure [37]. Since its release in 2010, the literature on GEE has grown exponentially, showing a wide range of vital applications including agriculture [38], wetlands [39], forests [40], and bathymetry [41], to name a few. GEE is free-to-use, and houses a number of EO datasets of various spatial and temporal resolutions, such as the popular Sentinel-1 and -2, Landsat, and MODIS archives.

Google Earth Engine (GEE) also provides access to numerous advanced ML algorithms and high-speed parallel processing [42], making the mapping and monitoring of environmental variables at local to global scales realizable. The built-in ML capabilities of GEE are user friendly, and allow for solving tasks such as supervised classification, unsupervised classification, and continuous regression problems [43]. Popular ML algorithms on GEE include Classification and Regression Trees (CART), Random Forest, and Support Vector Machine (SVM; [44]). These aforementioned algorithms represent some of the prominent ML techniques that have helped revolution EO, by helping process large and diverse datasets and determining relationships between spectral measurements and environmental phenomenon [45].

2. Materials and methods

This case study leveraged airborne and satellite RS, ML modeling, and cloud-based processing to estimate large-scale ALT. The general methodology is presented in **Figure 2**. The following sections detail each component, followed by the results of the study.

2.1 Study area

Active layer thickness (ALT) modeling and mapping was conducted across mainland Northwest Territories (NWT), Canada, covering an area of 114,000 km^2. Permafrost is found throughout all of the region. However, it is more common, thicker, and colder in the northern regions. This is represented by the three distinct permafrost zones which the NWT falls within: sporadic, discontinuous, and continuous. These zones are shown in **Figure 3**, along with the airborne RS data swaths which

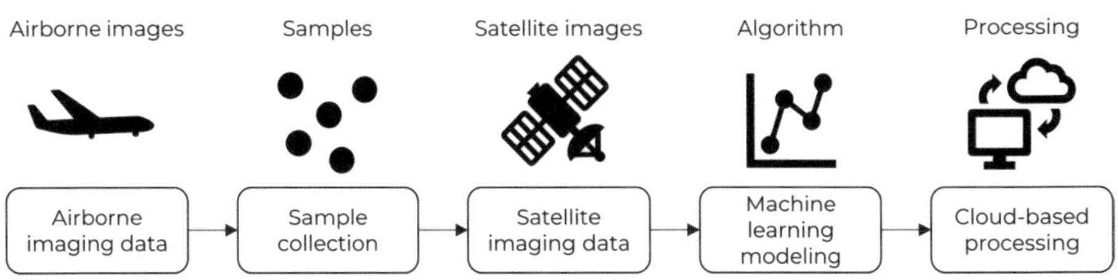

Figure 2.
Overall methodology used in this case study for ALT mapping.

are described in the forthcoming sections. Research is showing that permafrost conditions across the NWT are changing, with main drivers being increased air temperatures and changes to precipitation [46].

2.2 Airborne remote sensing data

Active layer thickness (ALT) measurements used for the development, optimization, and evaluation of a ML model were acquired via active airborne SAR imaging, by Chen et al. [47]. These data were collected during the 2017 NASA Arctic Boreal Vulnerability Experiment (ABoVE) airborne campaign, and therefore represent 2017 conditions. The ABoVE program, started in 2015, is a terrestrial ecology program researching environmental change across Alaska and Western Canada using RS [48]. The ALT profiles were retrieved using joint L- and P-band SAR sensors, from NASA's Uninhabited Aerial Vehicle Synthetic Aperture Radar (UAVSAR) and Airborne Microwave Observatory of Subcanopy and Subsurface (AirMOSS) SARs. As seen in **Figure 3**, 18 flight paths were conducted over the NWT, all of which were leveraged in this study. These ALT profiles have been independently validated by Parsekian et al. [49], and show good estimation compared to in situ GPR measurements.

2.3 Satellite earth observation data

The airborne RS data were upscaled to satellite EO data, since the latter are well-suited for regional mapping applications. All EO data were processed using GEE's JavaScript application programming interface (API). EO image sources included multi-frequency SAR, from the C-band Sentinel-1 and L-band ALOS PALSAR satellites, and electro-optical imagery from Landsat-8. These EO datasets were selected since SAR can penetrate vegetation and soil surfaces and interact with sub-surface hydrological conditions [33], and Landsat-8 supplies both reflectance

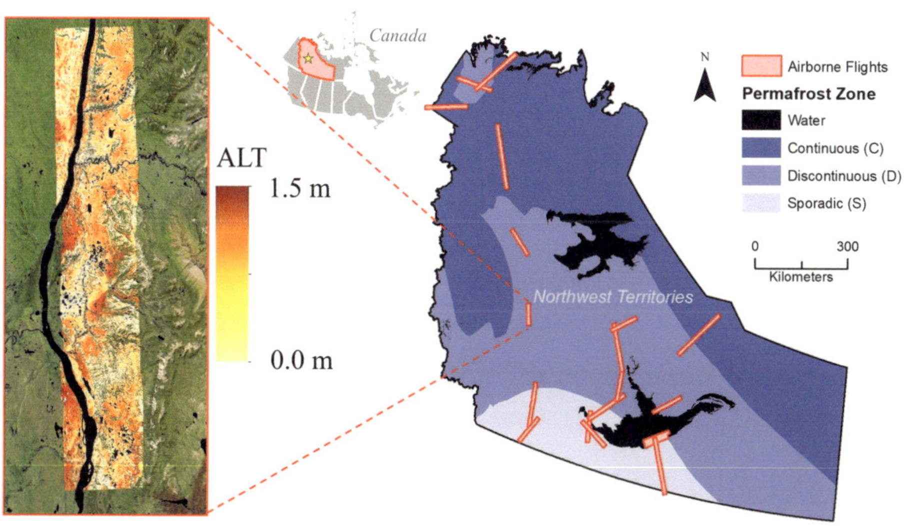

Figure 3.
Location of the study area, NWT, Canada. Extent of ALT measurements collected by NASA airborne flights during the ABoVE campaign are shown in red.

and thermal energy measurements which can help differentiate frozen from unfrozen ground [50]. EO image preprocessing followed that of Merchant et al. [39], and included fundamental steps such as speckle filtering and orthorectification of SAR, and cloud-masking of Landsat-8 imagery. High-resolution topographic imagery from the ArcticDEM project was also included in the modeling pipeline.

A time-series of images were collected for the SAR and electro-optical datasets, for the period of May-to-September 2017. These dates and range matched the airborne imaging acquisitions. The time-series for each EO source were then composed using a variety of aggregation functions in GEE (i.e., statistical descriptors; [51]), including mean, median, minimum, maximum, standard deviation, variance, and coefficient of variation. Each EO band and indice/variable were subject to these statistical functions, except for the time-static ArcticDEM variables. The list of satellite based EO variables used to model ALT across the NWT is found in **Table 1**.

2.4 Machine learning modeling

A ML approach was used to upscale the airborne ALT measurements to satellite observations. A Random Forest algorithm was chosen for this task, as it has shown to be robust and accurate for a variety of RS regression problems with GEE [52]. Moreover, it tends to handle non-linear and multidimensional geospatial data very well [53]. Random Forest is an ensemble ML algorithm that constructs multiple decision trees on random subsets of data, which then merges their predictions [54]. This improves regression performance and reduces model overfitting. Random forest has several hyperparameters that must be set in GEE. In order to optimize these values, a subset of ALT validation samples (which are described later) were used to tune each hyperparameter using the Scikit-learn and GridSearchCV packages in

EO source	Variable	Description	Statistical descriptors
Sentinel-1	σ°VV, σ°VH	C-band linear backscatter	Mean, median, minimum, maximum, standard deviation, coefficient of variation, variance
	VV/VH	C-band cross-pol ratio	
ALOS PALSAR	σ°HH, σ°HV	L-band linear backscatter	
	HH/HV	L-band cross-pol ratio	
Landsat-8	Blue, Green, Red, NIR, SWIR1, SWIR2	Surface reflectance bands	
	NDVI, NDWI	Normalized difference vegetation index and water index	
	LST	Land surface temperature	
	Albedo	Surface light fraction	
ArcticDEM	Elevation, slope, aspect	Topographic variables	N/A
	TPI	Topographic position index	

Table 1.
Satellite EO variables used for ALT estimation modeling. A total of 116 variables were collected.

Python. The tuning process used the k-fold statistical cross-validation method [55]. This method randomly divides the samples into k-groups (k = 5 was used) and then trains the model using k-1 folds. The ensuing model is then validated with the remaining samples.

The steps taken to train, validate (i.e., optimize), and test the ML modeling are shown in **Figure 4**. The first step involved sampling the airborne ALT profiles. A random stratification sampling approach was used for this. Two thousand five hundred randomly generated samples were first created within each 10 cm ALT bin, which ranged from 0 to 150 cm depth (e.g., 0 to 10 cm bin, 10 to 20 cm bin, and so on). This resulted in 37,500 sampling points, which was repeated four times to obtain four subsets of model training data, and then another two times to obtain model validation and testing data. The total number of sampling points is presented in **Table 2**. The training data subsets were used to train four separate ML models, with the final prediction result being the average ALT value. The validation samples were used to determine optimal hyperparameters for this modeling, in addition to performing

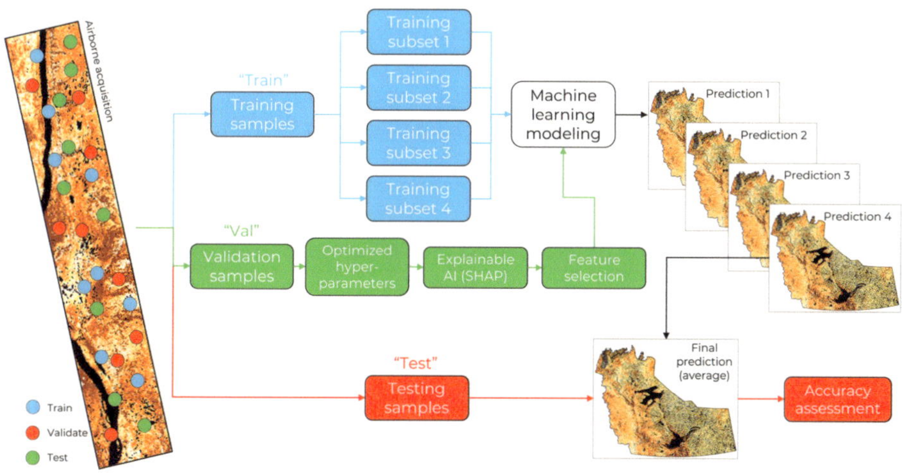

Figure 4.
Workflow for ML modeling. Blue: Data used for training. Green: Data used for validation. Red: Data used for testing.

Variable	Description	Number of samples
Training	Model 1	37,500
	Model 2	37,500
	Model 3	37,500
	Model 4	37,500
Validation		37,500
Testing		37,500
Total		225,000

Table 2.
Samples collected from airborne ALT profiles used to train, validate, and test the ML modeling.

a feature selection analysis, whereas the testing samples were used solely for final accuracy assessment.

2.5 Feature selection and model interpretation using explainable AI

Explainable AI (XAI) refers to a set of approaches used to explain and interpret ML models [56]. From these, the Shapley additive explanations (SHAP) algorithm was chosen to interpret the ML modeling of ALT [57]. SHAP is a state-of-the-art explainable AI (EAI) method, which can improve model transparency by revealing feature importance [58]. The algorithm borrows from cooperative game theory, in which the Shapley values quantify the additive, magnitudinal contribution of all variables to a model's prediction. Hence, SHAP was used to identify and rank the contributions of all 116 EO variables using the validation samples (**Table 2**), in order to support a feature selection process. The 30 most important EO features identified by the SHAP algorithm were chosen as final inputs to the Random Forest modeling.

2.6 Model performance assessment

Several statistical metrics were used to assess ML model performance, including coefficient of determination (R^2), room mean square error (RMSE), mean absolute error (MAE), mean square error (MSE), mean bias error (MBE), and relative absolute error (RAE):

$$R^2 = \frac{\left(\sum_{i=1}^{n}(x_i - \overline{x})(y_i - \overline{y})\right)^2}{\sum_{i=1}^{n}(x_i - \overline{x})\sum_{i=1}^{n}(y_i - \overline{y})^2} \tag{1}$$

$$\mathrm{RMSE} = \sqrt{\frac{1}{n}\sum_{i=1}^{n}(y_i - x_i)^2} \tag{2}$$

$$MAE = \frac{1}{n}\sum_{i=1}^{n}|y_i - x_i| \tag{3}$$

$$MSE = \frac{1}{n}\sum_{i=1}^{n}(y_i - x_i)^2 \tag{4}$$

$$MBE = \frac{1}{n}\sum_{i=1}^{n}(x_i - y_i) \tag{5}$$

$$RAE = \frac{\sum_{i=1}^{n}|y_i - x_i|}{\sum_{i=1}^{n}|x_i - \overline{x}|} \tag{6}$$

where x_i is the measured ALT value, y_i is the predicted (i.e., modeled) ALT value, $\overline{x}$ is the mean of the measured ALT, $\overline{y}$ is the mean of the predicted ALT, and n is the total of x and y. Model performance is best when R^2 is high, MBE is zero, and the remaining metrics are smaller [59].

3. Study results and discussion

3.1 Model tuning results

The ML tuning process, which took place prior to model training and application on GEE and using the subset of validation samples, resulted in a set of optimal Random Forest hyperparameters. The optimized hyperparameters, presented in **Table 3**, were selected based on the best model performance from the *k*-fold cross-validation procedure. The achieved R^2 from the cross-validation process was 0.432.

3.2 Explainable AI results

The results of the SHAP XAI analysis are presented in **Figure 5**, which includes both global (5a) and local (5b) feature importance plots. In both plots, only the top 30 (of 116) features are displayed, and are ordered by decreasing model importance. This is because the larger the SHAP value, the greater the contribution. These 30 features were selected as inputs to the ML modeling on GEE.

The global plot represents the mean SHAP value for each EO variable, which is quantified using all samples and therefore represents the average impact on model predictions. It is considered a quick and easily interpretable representation of SHAP results. This figure indicates that elevation, Landsat-8 (L8) mean and minimum LST, TPI, and slope are the five most important features explaining ALT variability.

The average global SHAP plot seen in **Figure 5a** is straightforward, however it does omit important information on variable contributions, such as the influence (i.e., direction) of samples on model construction. Hence, the summary plot is presented in **Figure 5b**, which combines feature importance and effects. Every point in the summary plot is a SHAP value for a single instance of a single EO variable. The values indicate the probability of different predicted ALT depths, based on the positive or negative correlation to ALT. For example, examination of this plot shows that low elevations, low topographic positions, and low slopes have a strong association with deeper ALTs. Deeper ALTs are also associated with warmer LSTs (both average and minimum temperatures). These findings generally align with the literature, in terms of the linkages between topography, temperatures, and permafrost [60].

3.3 Earth observation variable analysis

The most important EO variables, defined by the SHAP algorithm, were further analysed prior to ML modeling on GEE. In **Figure 6**, a correlation matrix shows the

Hyperparameter	Values assessed	Optimal value
'n_estimators'	[25, 50, 75, 100]	100
'max_depth'	[None, 10, 20, 30]	None
'min_samples_split'	[2, 5, 10, 15]	2
'min_samples_leaf'	[1, 2, 4, 8]	2

Table 3.
Optimal Random Forest hyperparameter settings identified by the cross-validation procedure.

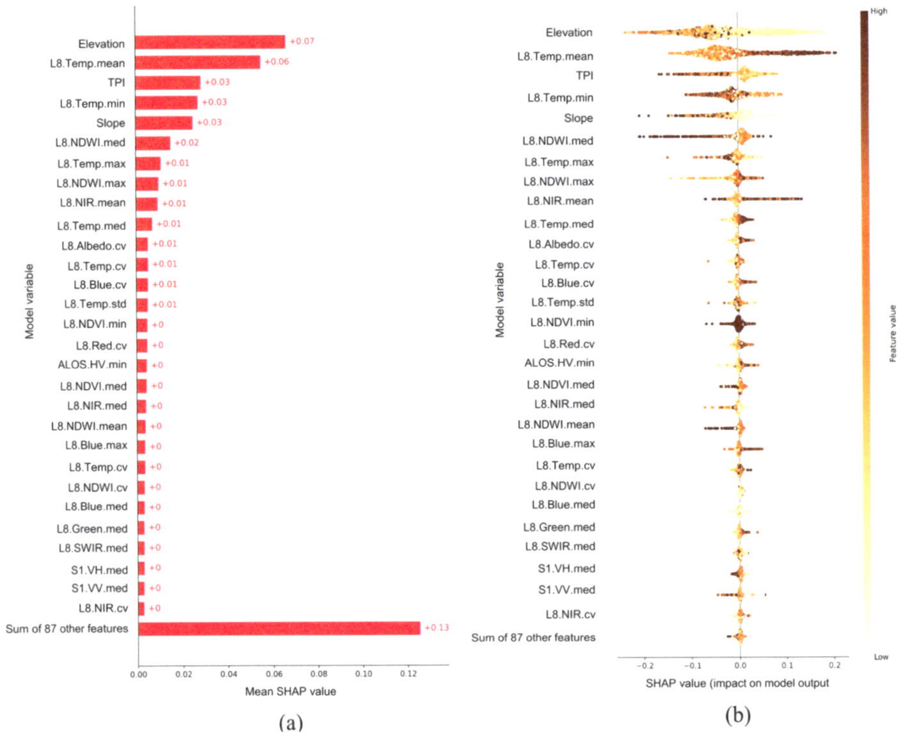

Figure 5.
(a) Global feature importance plot, displaying the mean absolute SHAP values, and (b) local feature importance plot, displayed using a SHAP beeswarm summary plot.

relationship between the 30 most important EO variables. Some variables showed strong correlations, such as between temperature and reflectance variables from L8. However, beyond these, most variables were not strongly correlated. This is important to note, because ML models tend to perform better when input variables are non-redundant, and multicollinearity (i.e., near-linear dependencies) is minimized [61].

Violin plots were also used to visualize the distribution and summary statistics of the 10 most important EO variables. In **Figure 7**, these plots were generated for shallow (0 to 0.75 m) and deep (0.75 to 1.5 m) ALTs, thus providing additional insight into the information captured by EO satellites. In particular, it can be seen that ALT characterization is relatively challenging for any single EO variable, which suggests that a multi-source fusion approach (i.e., SAR, optical, topography, etc.) is necessary for ALT estimation and mapping.

3.4 Active layer thickness modeling results

Following the tuning of hyperparameters and selection of important features, ML modeling of ALT was then implemented on the GEE. Statistical evaluation of the optimized ML model is found in **Table 4**. Model performance was also assessed by ecosystem type (e.g., wetlands) and vegetation (e.g., treed or shrub/grassland landcovers) based on the North American Land Change Monitoring (NALCM; [62]) landcover dataset. The ML model achieved an R^2 of 0.476 and RMSE of 0.313 m when evaluated across all landcovers. This performance was slightly weaker

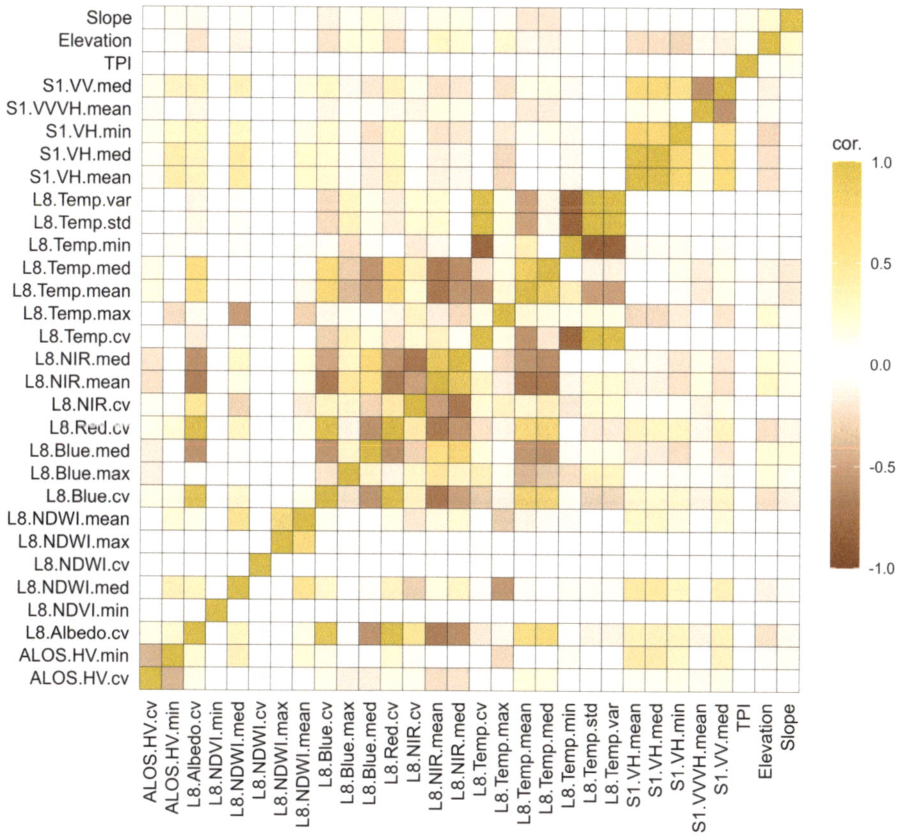

Figure 6.
Correlation matrix showing the relationship between the 30 most important EO variables.

in wetland ecosystems (R^2 of 0.456 and RMSE of 0.299 m) and in lower biomass areas with grassy/shrub coverage (R^2 of 0.433 and RMSE of 0.295 m), however it was stronger in treed ecosystems (R^2 of 0.497 and RMSE of 0.659 m). With wetlands, saturated conditions enhance heat transfer to layers below [63], resulting in deeper permafrost depths that may be more challenging to model with EO. In contrast, forested ecosystems provide canopy shading properties that inhibit radiation-induced thaw [64], leading to shallower ALT. These characteristics have been shown to relate strongly to EO data, such as those from optical sensors (e.g., NDVI; [27]).

Scatterplots are presented in **Figure 8** for additional visualization of model performance, which compares the airborne measured ALT to the ML predicted ALT. In general, shallower depths were better predicted than deeper depths, especially in wetlands which had deep ALTs.

3.5 Active layer thickness map of the Northwest Territories

The cloud-computing resources provided by GEE were used to scale the ML predictions across mainland NWT, covering an area of 114,000 km^2. The final results of the modeling are presented in the **Figure 9** map. Two areas are highlighted in this

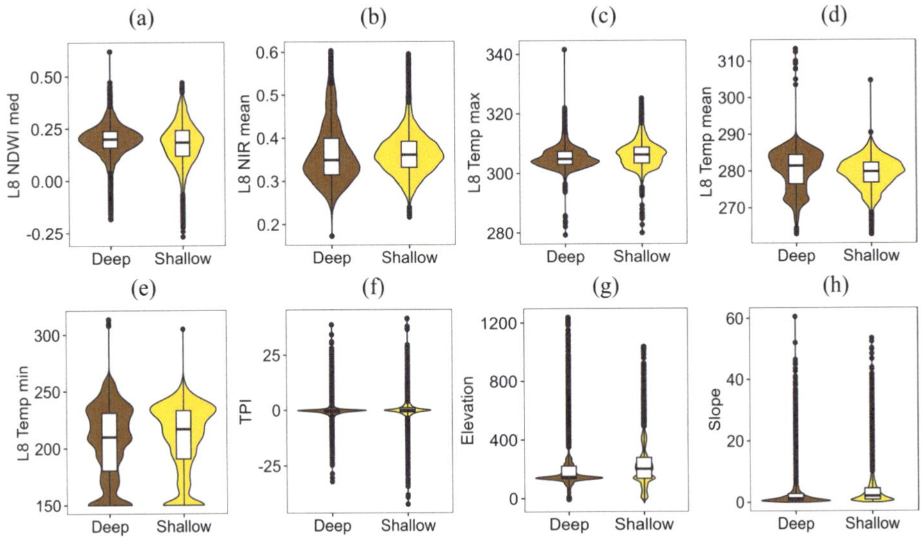

Figure 7.
Violin plots showing the distribution of pixels for the 10 most important EO variables. The plots show the dataset distribution and summaries by ALT depth, whereby "deep" corresponds to 0–0.75 m depths, and "shallow" corresponds to 0.75–1.5 m depths.

Landcover	R^2	MAE	MSE	RMSE	RAE	MBE
All	0.476	0.251	0.098	0.313	0.675	−0.003
Wetlands	0.456	0.234	0.089	0.299	0.678	0.002
Treed	0.497	0.264	0.106	0.327	0.659	−0.008
Grass/shrub	0.433	0.234	0.087	0.295	0.708	−0.0003

Table 4.
Performance evaluation of the ML model.

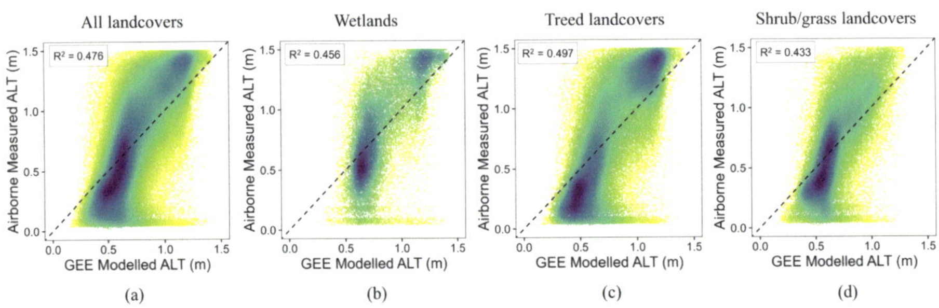

Figure 8.
Scatterplots of ALT predicted using ML modeling on GEE.

map, the Slave River Delta and the Tuktoyaktuk Coastlands. The former is a unique area in southern NWT consisting of vast wetland ecosystems with deep ALTs [66], whereas the latter is in the ice-rich high-latitudes where permafrost tends to be

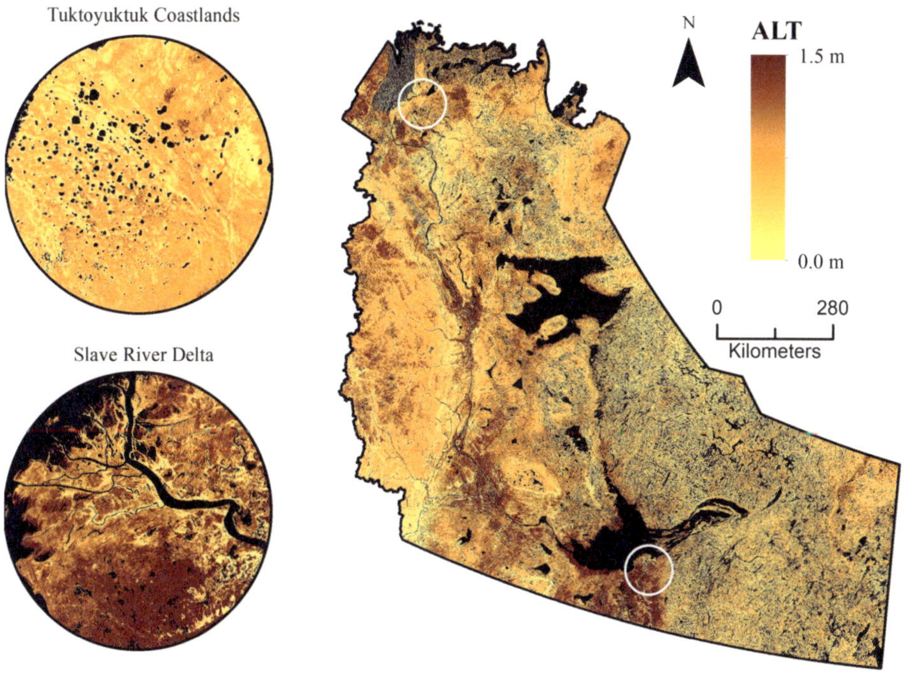

Figure 9.
ALT map of the NWT produced using ML on GEE. Water bodies were masked using data from the Global Surface Water Dynamics (GSWD; [65]) project for the year 2017.

thicker and closer to the surface [67]. For example, based on the modeling outputs, the sporadic permafrost zone of the NWT was found to have the highest percentage of deep ALTs (3.9%; **Figure 10**), whereas the discontinuous and continuous zones (i.e., those in the higher latitudes) had the highest percentage of shallow ALTs, respectively (i.e., thick permafrost close to the surface; 22.4 and 18.9%).

3.6 Advantages of cloud-based processing

Advanced geotechnologies were leveraged in this case study for ALT mapping across a rapidly changing landscape in Canada's Arctic region. GEE's computational resources and rich catalog of EO data, of which is Petabytes in scale, enabled the efficient access, preprocessing, and analysis of thousands of multi-source satellite images (**Figure 11**). In particular, dense time-series of RS big data were rapidly reduced (i.e., aggregated) into ready-to-analyse composites using an array of statistical functions (e.g., median, mean, etc.; **Figure 12**). Thus, the use of GEE removed the need for powerful and expensive local computing power (i.e., high-performance computing; HCP), which is otherwise necessary for scaling high-resolution predictions over large areas with RS big data [35]. This study now joins a dense collection of literature demonstrating the large-scale modeling capabilities of GEE for environmental applications and information generation over permafrost rich landscapes [33, 68–71]. As such, the use of GEE, along with other cloud-based geospatial platforms (e.g., Microsoft Azure, Amazon Web Services, etc.), is a continuously emerging trend in the field of RS and EO [72].

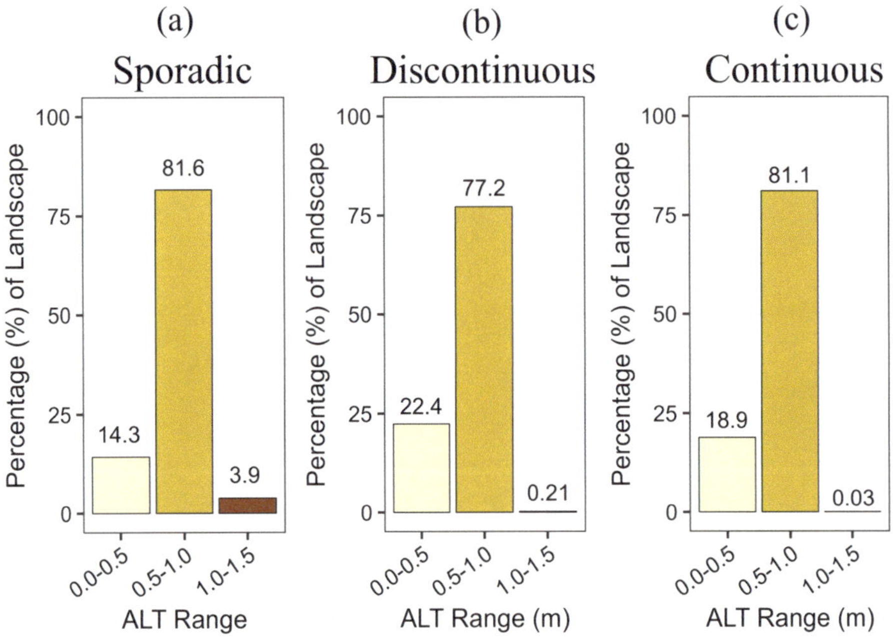

Figure 10.
Summary of ALT depths by permafrost zone across the NWT.

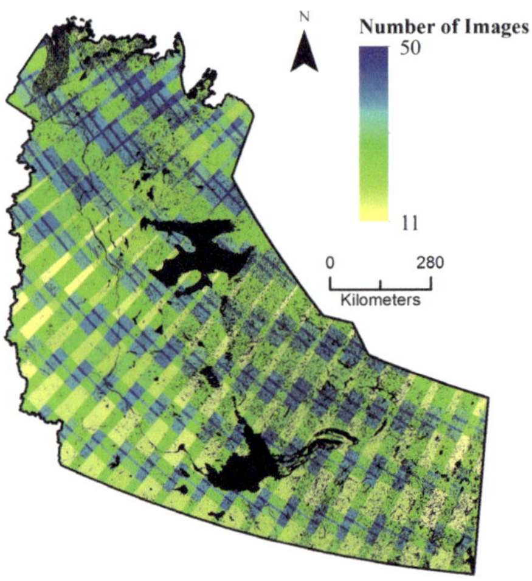

Figure 11.
Number of Landsat-8 satellite images processed over the study area for ALT estimation.

3.7 Implications for permafrost region science

Permafrost underlies much of the high-latitudes, and consequently it interacts with numerous systems including climate and human settlements [73], to name a few. Having the ability to map permafrost-related characteristics at high spatial resolutions, such as the ALT phenomenon, is important for better understanding the impacts on these and other critical systems [74]. This importance is magnified due to the accelerated warming of northern Arctic regions, which is stimulating permafrost

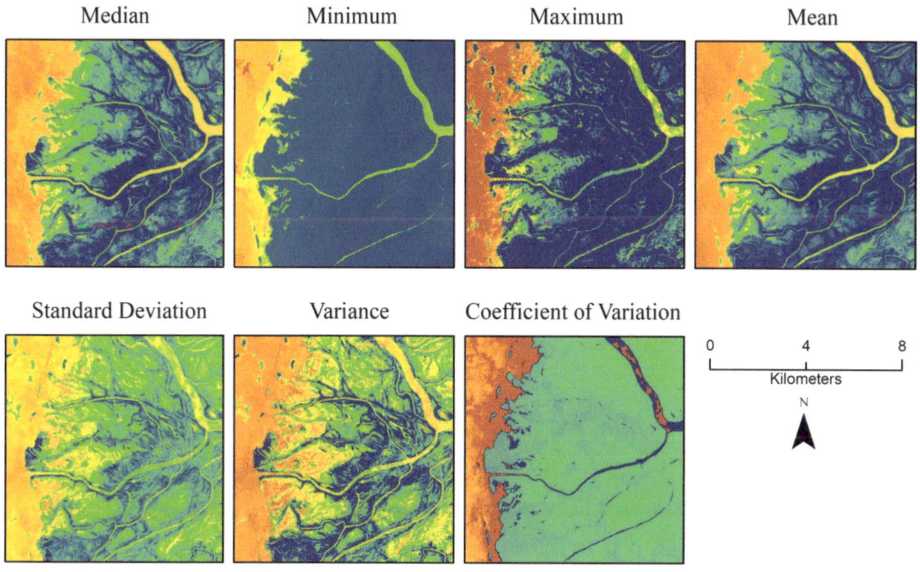

Figure 12.
Visual example of time-series statistical descriptors for the NDVI variable, over the Slave River Delta, NWT.

degradation [22]. For example, thawing of permafrost can induce slope instabilities [75] and destroy valuable infrastructure [7], while at the same time altering soil hydrology, nutrient, and microbial properties that impact vegetation [76]. Hence, there is a need for improved knowledge on the spatial distribution and characterization of the permafrost active layer, in order to support monitoring efforts and better understand the impact of a changing climate. The ALT map produced in this case study helps address these needs, by potentially serving as an essential input to climate and carbon storage models, land use planning processes, and baseline ecosystem analyses. Hence why many aspects of permafrost research require high-resolution spatial distribution knowledge on variables such as ALT, and the associated seasonal freezing patterns [19].

4. Conclusions

This chapter has explored the use of new EO technologies, with a particular focus on geospatial "big data" analysis with cloud-computing. By presenting a case study on the mapping of ALT, it has been demonstrated that these emerging technologies can be used to estimate challenging and heterogenous environmental variables, at scale. Analyzing such enormous time-series stacks of EO data was once a near-impossible task using conventional RS methods, and only reserved for those with super-computing resources. However, breakthroughs in cloud processing platforms, such as the GEE, have helped the RS community overcome challenges related to access, processing, and storage. The results of this case study highlight the capabilities of innovative EO technologies for permafrost environment mapping, by leveraging new RS sensors, cloud-processing, and ML to estimate a climatically significant variable, the ALT.

Author details

Michael A. Merchant* and Lindsay McBlane
Ducks Unlimited Canada, Edmonton, Alberta, Canada

*Address all correspondence to: michael.allan.merchant@gmail.com

References

[1] Zhang T, Heginbottom JA, Barry RG, Brown J. Further statistics on the distribution of permafrost and ground ice in the northern hemisphere. Polar Geography. 2000;**24**:126-131

[2] Luo D, Wu Q, Jin H, Marchenko SS, Lü L, Gao S. Recent changes in the active layer thickness across the northern hemisphere. Environment and Earth Science. 2016;**75**:555

[3] Throckmorton H, Newman B, Heikoop J, et al. Active layer hydrology in an Arctic tundra ecosystem: Quantifying water sources and cycling using water stable isotopes. Hydrological Processes. 2016;**30**:4972-4986

[4] Woo. Northern wetlands. In: Permafrost Hydrology. Berlin, Heidelberg: Springer; 2012. pp. 347-406

[5] Smith S, O'Neill H, Isaksen K, Noetzli J, Romanovsky V. The changing thermal state of permafrost. Nature Reviews Earth and Environment. 2022;**3**:10-23

[6] Li G, Zhang M, Pei W, Melnikov A, Khristoforov I, Li R, et al. Permafrost extent and active layer thickness variation in the northern hemisphere from 1969 to 2018. Science of the Total Environment. 2022;**804**:150182

[7] Hjort J, Streletskiy D, Doré G, Wu Q, Bjella K, Luoto M. Impacts of permafrost degradation on infrastructure. Nature Reviews Earth and Environment. 2022;**24**:3-1

[8] Schuur EAG, Mack MC. Ecological response to permafrost thaw and consequences for local and global ecosystem services. Annual Review of Ecology, Evolution, and Systematics. 2018;**49**:279-301

[9] Fisher JP, Estop-Aragonés C, Thierry A, Charman DJ, Wolfe SA, Hartley IP, et al. The influence of vegetation and soil characteristics on active-layer thickness of permafrost soils in boreal forest. Global Change Biology. 2016;**22**:3127-3140

[10] Wu Q, Hou Y, Yun H, Liu Y. Changes in active-layer thickness and near-surface permafrost between 2002 and 2012 in alpine ecosystems, Qinghai-Xizang (Tibet) plateau, China. Global and Planetary Change. 2015;**124**:149-155

[11] Chadburn S, Burke E, Cox P, Friedlingstein P, Hugelius G, Westermann S. An observation-based constraint on permafrost loss as a function of global warming. Nature Climate Change. 2017;7:340-344

[12] Walvoord MA, Kurylyk BL. Hydrologic impacts of thawing permafrost—A review. Vadose Zone Journal. 2016;**15**:1-20

[13] Chen L, Liang J, Qin S, Liu L, Fang K, Xu Y, et al. Determinants of carbon release from the active layer and permafrost deposits on the Tibetan plateau. Nature Communications. 2016;7:13046

[14] Schaefer K, Zhang T, Bruhwiler L, Barrett AP, Fe RAE, Ju NZHANGIG, et al. Amount and timing of permafrost carbon release in response to climate warming. Chemical and Physical Meteorology. 2011;**63**:165-180

[15] Schaefer K, Lantuit H, Romanovsky VE, Schuur EAG, Witt R. The impact of the permafrost carbon feedback on global climate. Environmental Research Letters. 2014;**9**:9

[16] Miner K, Turetsky M, Malina E, Bartsch A, Tamminen J, McGuire AD, et al. Permafrost carbon emissions in a changing Arctic. Nature Reviews Earth and Environment. 2022;**3**:55-67

[17] Rantanen M, Karpechko AY, Lipponen A, Nordling K, Hyvärinen O, Ruosteenoja K, et al. The Arctic has warmed nearly four times faster than the globe since 1979. Communications Earth & Environment. 2022;**3**:168

[18] You Q, Cai Z, Pepin N, et al. Warming amplification over the Arctic pole and third pole: Trends, mechanisms and consequences. Earth-Science Reviews. 2021;**217**:103625

[19] Dobiński W. Permafrost active layer. Earth-Science Reviews. 2020;**208**:103301

[20] Sudakova M, Sadurtdinov M, Skvortsov A, Tsarev A, Malkova G, Molokitina N, et al. Using ground penetrating radar for permafrost monitoring from 2015-2017 at calm sites in the Pechora river delta. Remote Sensing. 2021;**13**:3271

[21] Yi Y, Kimball JS, Chen RH, Moghaddam M, Reichle RH, Mishra U, et al. Characterizing permafrost active layer dynamics and sensitivity to landscape spatial heterogeneity in Alaska. The Cryosphere. 2018;**12**:145-161

[22] Jorgenson MT, Grosse G. Remote sensing of landscape change in permafrost regions. Permafrost and Periglacial Processes. 2016;**27**:324-338

[23] Philipp M, Dietz A, Buchelt S, Kuenzer C. Trends in satellite earth observation for permafrost related analyses—A review. Remote Sensing. 2021;**13**:1217

[24] Pastick N, Jorgenson M, Wylie B, Minsley B, Ji L, Walvoord M, et al. Extending airborne electromagnetic surveys for regional active layer and permafrost mapping with remote sensing and ancillary data, Yukon flats ecoregion, Central Alaska. Permafrost and Periglacial Processes. 2013;**24**:184-199

[25] Cao H, Gao B, Gong T, Wang B. Analyzing changes in frozen soil in the source region of the yellow river using the Modis land surface temperature products. Remote Sensing. 2021;**13**:180

[26] Zhang C, Douglas TA, Anderson JE. Modeling and mapping permafrost active layer thickness using field measurements and remote sensing techniques. International Journal of Applied Earth Observation and Geoinformation. 2021;**102**:102455

[27] Luo D, Liu L, Jin H, Wang X, Chen F. Characteristics of ground surface temperature at Chalaping in the source area of the yellow river, northeastern Tibetan plateau. Agricultural and Forest Meteorology. 2020;**281**:107819

[28] Xu X, Wu Q. Active layer thickness variation on the Qinghai-Tibetan plateau: Historical and projected trends. Journal of Geophysical Research: Atmospheres. 2021;**126**:e2021JD034841

[29] Zhan D, Li M, Xiao Y, Man H, Zang S. Spatial differentiation and influencing factors of active layer thickness in the Da Hinggan Ling prefecture. Frontiers in Earth Science. 2023;**10**:1066662

[30] Merchant M, Adams J, Berg A, Baltzer J, Quinton W, Chasmer L. Contributions of C-band SAR data and polarimetric decompositions to subarctic boreal peatland mapping. IEEE Journal of Selected Topics in Applied Earth Observations and Remote Sensing. 2017;**10**:1467-1482

[31] Widhalm B, Bartsch A, Leibman M, Khomutov A. Active-layer thickness

estimation from X-band SAR backscatter intensity. The Cryosphere. 2017;**11**:483-496

[32] Wang C, Zhang Z, Zhang H, Zhang B, Tang Y, Wu Q. Active layer thickness retrieval of Qinghai-Tibet permafrost using the TerraSAR-X InSAR technique. IEEE Journal of Selected Topics in Applied Earth Observations and Remote Sensing. 2018;**11**:4403-4413

[33] Merchant M, Obadia M, Brisco B, Devries B, Berg A. Applying machine learning and time-series analysis on Sentinel-1A SAR/InSAR for characterizing Arctic tundra hydro-ecological conditions. Remote Sensing. 2022;**14**:1123

[34] Zhang X, Zhang H, Wang C, Tang Y, Zhang B, Wu F, et al. Active layer thickness retrieval over the Qinghai-Tibet plateau using Sentinel-1 multitemporal InSAR monitored permafrost subsidence and temporal-spatial multilayer soil moisture data. IEEE Access. 2020;**8**:84336-84351

[35] Ma Y, Wu H, Wang L, Huang B, Ranjan R, Zomaya A, et al. Remote sensing big data computing: Challenges and opportunities. Future Generation Computer Systems. 2015;**51**:47-60

[36] Gomes VCF, Queiroz GR, Ferreira KR. An overview of platforms for big earth observation data management and analysis. Remote Sensing. 2020;**12**:1253

[37] Tamiminia H, Salehi B, Mahdianpari M, Quackenbush L, Adeli S, Brisco B. Google Earth Engine for geo-big data applications: A meta-analysis and systematic review. ISPRS Journal of Photogrammetry and Remote Sensing. 2020;**164**:152-170

[38] Phalke AR, Özdoğan M, Thenkabail PS, Erickson T, Gorelick N, Yadav K, et al. Mapping croplands of Europe, Middle East, Russia, and Central Asia using Landsat, Random forest, and Google Earth Engine. ISPRS Journal of Photogrammetry and Remote Sensing. 2020;**167**:104-122

[39] Merchant M, Brisco B, Mahdianpari M, Bourgeau-Chavez L, Murnaghan K, DeVries B, et al. Leveraging Google Earth Engine cloud computing for large-scale Arctic wetland mapping. International Journal of Applied Earth Observation and Geoinformation. 2023;**125**:103589

[40] Jahromi M, Zolghadr-Asli B, Pourghasemi H, Alavipanah S. Google Earth Engine and its application in forest sciences. In: Spatial Modeling in Forest Resources Management: Rural Livelihood and Sustainable Development. New York City, USA: Springer; 2021. pp. 629-649

[41] Merchant MA. Modelling inland Arctic bathymetry from space using cloud-based machine learning and Sentinel-2. Advances in Space Research. 2023;**72**:4256-4271

[42] Yang L, Driscol J, Sarigai S, Wu Q, Chen H, Lippitt CD. Google Earth Engine and artificial intelligence (AI): A comprehensive review. Remote Sensing. 2022;**14**:1-110

[43] Gorelick N, Hancher M, Dixon M, Ilyushchenko S, Thau D, Moore R. Google Earth Engine: Planetary-scale geospatial analysis for everyone. Remote Sensing of Environment. 2017;**202**:18-27

[44] Feizizadeh B, Omarzadeh D, Kazemi Garajeh M, Lakes T, Blaschke T. Machine learning data-driven approaches for land use/cover mapping and trend analysis using Google Earth Engine. Journal of Environmental Planning and Management. 2021;**66**:665-697

[45] Maxwell AE, Warner TA, Fang F. Implementation of machine-learning classification in remote sensing: An applied review. International Journal of Remote Sensing. 2018;**39**:2784-2817

[46] Mamet SD, Chun KP, Kershaw GGL, Loranty MM, Peter Kershaw G. Recent increases in permafrost thaw rates and areal loss of palsas in the western northwest territories, Canada. Permafrost and Periglacial Processes. 2017;**28**:619-633

[47] Chen H, Michaelides R, Chen J, et al. ABoVE: Active Layer Thickness from Airborne L-and P-band SAR, Alaska, 2017, Ver. 3. Oak Ridge, Tennessee, USA: ORNL DAAC; 2022

[48] Miller CE, Griffith PC, Goetz SJ, et al. An overview of above airborne campaign data acquisitions and science opportunities. Environmental Research Letters. 2019;**14**:080201

[49] Parsekian AD, Chen RH, Michaelides RJ, et al. Validation of permafrost active layer estimates from airborne SAR observations. Remote Sensing. 2021;**13**:2876

[50] Kalinicheva SV, Shestakova AA. Using thermal remote sensing in the classification of mountain permafrost landscapes. Journal of Mountain Science. 2021;**18**:635-645

[51] Zhang Z, Wei M, Pu D, He G, Wang G, Long T. Assessment of annual composite images obtained by google Earth Engine for urban areas mapping using random forest. Remote Sensing. 2021;**13**:1-19

[52] Belgiu M, Dragut L. Random forest in remote sensing: A review of applications and future directions. ISPRS Journal of Photogrammetry and Remote Sensing. 2016;**114**:24-31

[53] Merchant M, Haas C, Schroder J, Warren R, Edwards R. High-latitude wetland mapping using multidate and multisensor earth observation data: A case study in the northwest territories. Journal of Applied Remote Sensing. 2020;**14**:1-18

[54] Breiman L. Random forests. Machine Learning. 2001;**45**:5-32

[55] Kohavi R. A study of cross-validation and bootstrap for accuracy estimation and model selection. IJCAI. 1995;**14**:1137-1145

[56] Minh D, Wang HX, Li YF, Nguyen TN. Explainable artificial intelligence: A comprehensive review. Artificial Intelligence Review. 2022;**55**:3503-3568

[57] Shapley L. Value for n-person games. In: Kuhn H, Tucker A, editors. Contribution to the Theory of Games II. Princeton, NJ, USA: Princeton University Press; 1953. pp. 307-317

[58] Temenos A, Temenos N, Kaselimi M, Doulamis A, Doulamis N. Interpretable deep learning framework for land use and land cover classification in remote sensing using SHAP. IEEE Geoscience and Remote Sensing Letters. 2023;**20**:8500105

[59] Maulud DH, Abdulazeez AM. A review on linear regression comprehensive in machine learning. Journal of Applied Science and Technology Trends. 2020;**1**:140-147

[60] Zhang Y, Touzi R, Feng W, Hong G, Lantz TC, Kokelj SV. Landscape-scale variations in near-surface soil temperature and active-layer thickness: Implications for high-resolution permafrost mapping. Permafrost and Periglacial Processes. 2021;**32**:627-640

[61] Merchant MA, Warren RK, Edwards R, Kenyon JK. An object-based assessment of multi-wavelength SAR, optical imagery and topographical datasets for operational wetland mapping in boreal Yukon, Canada. Canadian Journal of Remote Sensing. 2019;**45**:308-332

[62] Giri C. Remote Sensing of Land Use and Land Cover: Principles and Applications. Boca Raton, Florida, USA: CRC Press; 2012

[63] Douglas TA, Turetsky MR, Koven CD. Increased rainfall stimulates permafrost thaw across a variety of interior Alaskan boreal ecosystems. NPJ Climate and Atmospheric Science. 2020;**3**:28

[64] Maria Stuenzi S, Boike J, Cable W, Herzschuh U, Kruse S, Pestryakova LA, et al. Variability of the surface energy balance in permafrost-underlain boreal forest. Biogeosciences. 2021;**18**:343-365

[65] Pickens AH, Hansen MC, Hancher M, Stehman SV, Tyukavina A, Potapov P, et al. Mapping and sampling to characterize global inland water dynamics from 1999 to 2018 with full Landsat time-series. Remote Sensing of Environment. 2020;**243**:111792

[66] Zhang F, Mosaffa M, Chu T, Lindenschmidt KE. Using remote sensing data to parameterize ice jam modeling for a northern inland delta. Water (Switzerland). 2017;**9**:306

[67] Westermann S, Duguay C, Grosse G, Kääb A. Remote sensing of permafrost and frozen ground. In: Remote Sensing of the Cryosphere. New Jersey, USA: John Wiley & Sons, Ltd.; 2015. pp. 307-344

[68] Nyland KE, Gunn GE, Shiklomanov NI, Engstrom RN, Streletskiy DA. Land cover change in the lower Yenisei river using dense stacking of Landsat imagery in Google Earth Engine. Remote Sensing. 2018;**10**:1226

[69] Pointner G, Bartsch A. Mapping arctic lake ice backscatter anomalies using Sentinel-1 time series on Google Earth Engine. Remote Sensing. 2021;**13**:1626

[70] Qi Y, Li S, Ran Y, Wang H, Wu J, Lian X, et al. Mapping frozen ground in the Qilian mountains in 2004-2019 using Google Earth Engine cloud computing. Remote Sensing. 2021;**13**:1-17

[71] Zakharov M, Gadal S, Kamicaityte J, Cherosov M, Troeva E. Distribution and structure analysis of mountain permafrost. Land. 2022;**11**:1-21

[72] Loukili Y, Lakhrissi Y, Ben ASE. Geospatial big data platforms: A comprehensive review. KN Journal of Cartography and Geographic Information. 2022;**72**:293-308

[73] Vecellio DJ, Nowotarski CJ, Frauenfeld OW. The role of permafrost in Eurasian land-atmosphere interactions. Journal of Geophysical Research: Atmospheres. 2019;**124**:11644-11660

[74] Gruber S. Derivation and analysis of a high-resolution estimate of global permafrost zonation. The Cryosphere. 2012;**6**:221-233

[75] Deline P, Gruber S, Amann F, et al. Ice loss from glaciers and permafrost and related slope instability in high-mountain regions. In: Snow and Ice-Related Hazards, Risks, and Disasters. Amsterdam, Netherlands: Elsevier; 2021. pp. 501-540

[76] Jin XY, Jin HJ, Iwahana G, Marchenko SS, Luo DL, Li XY, et al. Impacts of climate-induced permafrost degradation on vegetation: A review. Advances in Climate Change Research. 2021;**12**:29-47

Chapter 3

Advanced Digital Image Analysis of Remotely Sensed Data Using JavaScript API and Google Earth Engine

Anwarelsadat Elmahal and Eltaib Ganwa

Abstract

Earth observation satellites (EOS) have been continuously observing and monitoring the globe and its physical, chemical, and biological characteristics since the late 1950s. EOS has been providing an enormous quantity of spatial data, and it is predicted that the global satellite data market will expand exponentially. Processing such vast and unique data in an efficient manner is necessary. Google Earth Engine (GEE) is an example of a cloud-based geospatial platform that revolutionized the way users store, process, analyze, and visualize spatial data. This chapter delves into the fundamentals of digital image processing, employing cutting-edge techniques to process, analyze, and visualize remotely sensed data through the use of Google Earth Engine (GEE) and JavaScript API. This chapter presents an in-depth exploration of digital image processing, specifically focusing on remotely sensed data. It highlights advanced techniques and includes two case studies that demonstrate practical applications in preprocessing and detailed analysis of land use and land cover (LULC) changes.

Keywords: image preprocessing, image analysis, GEE, JavaScript API, LULCC

1. Introduction

The first artificial satellite, Sputnik 1, was launched by the Soviet Union on October 4, 1957 [1, 2]. Since then, the number of space missions has increased and Earth observation satellites (EOS) have been continuously observing and monitoring the globe and its physical, chemical, and biological characteristics [3, 4]. Satellite images have offered and continue to provide significant and reliable sources of information, this is because of their synoptic view, repetitive global coverage, and map-like format [5]. Common digital image processing techniques for information extraction have become inefficient as the number of satellites has increased significantly, along with the ongoing advances in spatial, spectral, and temporal data acquisition [5]. Therefore, we require new techniques to analyze remotely sensed data for a variety of reasons, including (i) improving data processing accuracy and efficiency; (ii) extracting new information with greater accuracy; and (iii) making data more accessible and usable [6–8]. Advances in the internet and technology have contributed to novel approaches in processing remotely sensed data in a variety of ways, including (i) increased data accessibility; (ii) increased computational power; and most importantly ; and (iii) the development of new algorithms [9, 10]. As a result, many online platforms for processing remotely sensed data have been created, such as Microsoft Planetary Computer; Earth on AWS; EOS Data Hub, and Google Earth Engine (GEE). However, Google Earth Engine is a leading online platform for processing remotely sensed data. It is more mature, has more features, and is better suited for advanced processing tasks than other platforms [9]. GEE is an online platform that leverages Google's multi-petabyte data catalog for analyzing environmental data at the planetary scale [9]. GEE as a spatial cloud computing platform is categorized as PaaS, a platform as a service [11].

In comparison to traditional on-premises deployment [12, 13], PaaS offers several advantages, such as (i) scalability, which eliminates the need to manage the underlying infrastructure, (ii) agility, which facilitates the quick and easy development and deployment of applications, and (iii) cost savings, because you only pay for the resources that you use. GEE is a powerful tool for digital image processing, and its code editor provides a structured environment for writing and debugging code, using either native or non-native. Users could access Google Earth Engine (GEE) via the JavaScript API, Python, and rgee, as outlined in reference [11]. Nonetheless, this chapter will focus specifically on the detailed exploration of the JavaScript API.

The chapter will introduce you to advanced digital image processing for remotely sensed data using the computational power of Google Earth Engine (GEE) and its native JavaScript (JS) API. The reader will be also introduced to the conceptual approaches in the areas of digital imaging systems, preprocessing, and visualizations of images. In conclusion, two practical case studies were implemented to offer users an interactive experience in learning how to effectively use GEE and the JS API. These hands-on exercises encompass a range of techniques, such as preprocessing, visualization, and image classification. Moreover, Case study 2 provides an in-depth analysis of image classification, focusing on key measures, such as overall accuracy and accuracy per class, offering comprehensive insights into the application of these processes in practical scenarios.

2. Basics of satellite imagery

Satellite imagery, enhanced by sophisticated sensors and instruments, captures electromagnetic radiation (EMR) that is either emitted or reflected from the Earth's surface, including the oceans and the atmosphere [14]. This includes the use of multispectral and hyperspectral images, across various wavelengths. Such technology provides detailed and nuanced information about the Earth's diverse features, including landforms, water bodies, and atmospheric conditions. The collected data provide an essential understanding of the planet's physical and environmental characteristics, covering a wide range of natural phenomena.

2.1 Remote sensing mechanisms

Based on their sensing mechanisms, satellites could be categorized into passive and active satellites [15]. Passive satellites operate by detecting, receiving, and registering

EMR that is either reflected or emitted from the Earth [16]. These satellites are outfitted with sensors capable of detecting different types of electromagnetic radiation (EMR), including visible and near-infrared, enabling them to produce clear and detailed images of the Earth's surface. Some examples of passive remote sensing satellites are the Landsat series, MODIS satellites, the Sentinel-2 mission, and the advanced very high resolution radiometer (AVHR). In contrast, active sensors like the synthetic aperture radar (SAR) on Sentinel-1 emit their own energy and analyze the reflected signal [16, 17]. This technology excels in terrain mapping and environmental monitoring under any weather and environmental conditions.

2.2 Image acquisition

Satellite remote sensing involves collecting electromagnetic radiation from the Earth or other celestial bodies using satellites, which are then converted into digital images. This process is characterized by several essential steps [15], which may include (i) the use of onboard sensors to capture the EMR; (ii) the transmission of this data to ground stations via high-frequency radio waves; and (iii) the processing of this data, which includes correcting or reducing errors and distortions and enhancing their overall quality for further in-depth analysis. Each stage is crucial, utilizing advanced technology and methods to guarantee the final images' precision and utility.

3. Digital image processing

Typically, data obtained through remote sensing is susceptible to both random and systematic errors, which can introduce noise and bias, thereby affecting the precision of the measurements [14, 16]. For numerous tasks related to images, such as computer vision, image analysis, and machine learning, image preprocessing serves as a vital initial step. In the context of remote sensing, digital image processing encompasses the use of computer algorithms for the manipulation and analysis of digital images. It includes tasks such as enhancing, restoring, and extracting valuable information from images, which is essential for a wide range of applications.

3.1 Challenges in satellite image processing

Satellite image processing entails overcoming various challenges, including atmospheric distortions, the need for significant storage and computational resources, and the impact of satellite movement on image consistency. Advanced correction techniques, precise calibration, and sophisticated analysis are essential to address these issues and unlock the full potential of satellite imagery for diverse applications.

3.2 Digital image processing steps

The main operations of remote sensing digital image processing and analysis could be categorized into three groups: (i) data preparation and enhancement, (ii) information extraction and analysis, and (iii) data integration and presentation. Refining and reducing errors in data obtained from remote sensing is essential for improving the accuracy and reliability of the data. Such enhancements help users across various fields to make better-informed choices.

3.2.1 Data preparation and enhancement

3.2.1.1 Image rectifications

Image distortions in remotely sensed data can result from sensor orientation, Earth's curvature, atmospheric conditions, and satellite dynamics [18]. The process of image rectification or preprocessing addresses both systematic and random errors acquired during data collection. Systematic errors, being consistent, can be corrected, whereas random errors are unpredictable and vary with each measurement [19, 20]. The necessity for error correction is determined by the error's significance, its impact on the analysis, the risk of introducing new errors, and the error pattern's consistency [14–16, 21]. The rectification process includes three key steps: (1) image preprocessing to correct sensor orientation and adjust for Earth's curvature and satellite dynamics; (2) geometric transformation and resampling to mathematically transform and interpolate pixel values for a specific map projection; and (3) radiometric and quality control to offset atmospheric effects and ensure image quality through radiometric calibration. These steps ensure that the rectified images are accurate and of high quality for remote sensing analyses.

3.2.1.2 Image registration

Image registration in remote sensing is a comprehensive process that aligns multiple satellite or aerial images of the same area taken at different times or angles and integrates these images with other data types, such as vector maps [22]. This integration provides an accurate and consistent representation of ground features, enhancing spatial accuracy and context by combining imagery with GIS maps. This approach is crucial for applications where accurate geographic information is key. The process involves feature detection, matching control points, applying a transformation model, and resampling the images with quality assurance measures.

Resampling methods adjust pixel values during image transformations, with the main types being nearest neighbor, bilinear interpolation, and bicubic interpolation. These methods strike a balance between image quality and computational efficiency, which is crucial for image scaling and geometric corrections.

Orthorectification corrects geometric distortions in aerial or satellite imagery, aligning the image to a standard map projection while accounting for Earth's curvature, camera tilt, and landscape topography [15, 16]. This ensures each pixel is accurately aligned to its true geographic position.

3.2.1.3 Atmospheric correction

The objective of the atmospheric correction in satellite and aerial imagery processing is to remove the distortions caused by the Earth's atmosphere [15]. Atmospheric correction is essential in remote sensing to ensure data accurately reflects the Earth's surface by adjusting for light scattering and absorption by atmospheric particles and gasses. This step strengthens the accuracy and reliability of the data, critical for environmental monitoring and LULC mapping. Methods to mitigate atmospheric distortions include dark object subtraction, atmospheric modeling, and radiative transfer models [14–16, 23]. This provides users with the flexibility to either apply these corrections or select images with minimal distortion for their work.

DOI: http://dx.doi.org/10.5772/intechopen.1004501

3.2.1.4 Noise reduction

Noise reduction in image processing is a technique used to minimize unwanted random variations, or "noise," in digital images. This noise, often seen as grainy specks, can degrade image quality. Noise removal could be considered as an enhancement procedure [16]. Noise reduction algorithms aim to smooth out these imperfections while preserving important details. Though essential in image processing, excessive noise reduction may cause a loss of detail, making it important to strike a balance.

3.2.1.5 Speckle filtering

Speckle filtering is a noise reduction technique used in radar imagery, especially for synthetic aperture radar (SAR) data, to mitigate the granular "speckle" noise. This noise, appearing as a salt-and-pepper pattern, hinders image clarity. Speckle filtering algorithms focus on reducing this noise while preserving essential image details [24]. It is crucial for applications such as environmental monitoring and agricultural mapping, where clear radar imagery is essential. However, careful application is needed to avoid losing important features in the process of noise reduction.

3.2.1.6 Gaussian smoothing

Gaussian smoothing is an image processing technique that uses a Gaussian filter to blur an image and reduce noise [18]. This method averages pixel values, giving more weight to central pixels, to create a subtle, natural-looking blur. It is commonly used as a preprocessing step to minimize noise before tasks like edge detection, helping to avoid false edges. However, it is important to apply this technique carefully to prevent excessive blurring and loss of important image details.

4. Image enhancement techniques

In the context of remote sensing, "image enhancement" refers to the process of improving the visual appearance of an image or making certain features of an image more discernible for analysis [15, 16]. This process is crucial in remote sensing as it helps in better interpreting and analyzing the data collected from satellite or aerial imagery. Image enhancement techniques are typically grouped into two main categories: spatial domain and frequency domain [16].

1. Spatial domain techniques, which could be further categorized into (i) contrast enhancement, (ii) spatial filtering, (iii) histogram equalizer, and (iv) edge filtering.

2. Frequency domain techniques, which could also be categorized into (i) Fourier Transform and (ii) frequency filtering.

4.1 Spatial domain techniques

These techniques operate directly on the pixels of the image. They are primarily concerned with modifying the pixel values in an image to achieve the desired enhancement.

4.1.1 Contrast enhancement

Contrast enhancement refers to the modification of an image's tonal distribution. This process entails scientifically adjusting pixel brightness values to either expand or compress the intensity spectrum. Such manipulation augments the visual distinction of different features within the image, thereby facilitating more effective analysis and interpretation for remote sensing applications.

4.1.2 Spatial filtering

Spatial filtering in digital image processing is a technique used to manipulate or enhance an image by altering the spatial frequency components. It involves applying filters, typically in the form of a kernel or matrix, over an image to perform operations such as smoothing, sharpening, and edge detection [25]. Smoothing filters reduce noise and detail, while sharpening filters enhance edges and fine details. Edge detection filters identify and outline significant transitions in intensity. Spatial filtering is pivotal in various applications, including enhancing image features for analysis, preprocessing for further image-processing tasks, and improving the visual quality of images.

4.1.3 Histogram equalizer

Histogram equalization is a method in image processing that adjusts the contrast of an image. It works by effectively spreading out the most frequent intensity values, thus stretching the range of intensity levels in the image's histogram. This results in an image with a more uniform distribution of intensities, enhancing the overall contrast [18]. Histogram equalization is particularly useful in images with backgrounds and foregrounds that are both bright or dark, making it easier to identify features and details in such images. Histogram equalization can sometimes overly enhance noise, obscure details in very light or dark areas, and may not be suitable for every kind of imagery.

4.1.4 Edge filtering

The objective of edge filtering is to detect and enhance edges within an image. Edges in an image are significant local changes in intensity, typically occurring at the boundaries of different objects or features [26]. Edge filtering is accomplished by applying various algorithms or operators, which work by highlighting intensity transitions [27]. These filters operate by convolving a kernel or filter matrix over the image. This process calculates a new value for each pixel by combining it with its neighboring pixel values, effectively emphasizing areas of high-intensity change. The result is an image where the edges are more pronounced, making it easier to identify shapes, boundaries, and structural details in the image [27].

Edge filtering is crucial in various applications, including image analysis, computer vision, pattern recognition, and machine learning, as it aids in object detection, segmentation, and feature extraction. Edge filtering in remote sensing is crucial for many applications to list a few, precision agriculture, environmental monitoring, urban planning, road mapping, geological studies, and archeological exploration.

4.2 Frequency domain techniques

The frequency domain, crucial in signal processing and related fields, focuses on analyzing signals or mathematical functions by their frequency components rather than time-based behavior. This may involve transforming an image into its sinusoidal frequency components, typically using mathematical methods like the Fourier Transform [22]. In this domain, different frequencies represent various image features, with low frequencies indicating gradual intensity changes and high frequencies highlighting rapid changes such as edges and details. This approach is advantageous for tasks such as filtering, image compression, and frequency-based image enhancement, as it allows more efficient and effective processing.

4.2.1 Fourier Transform

The Fourier Transform (FT) is a mathematical transform widely used in signal processing and physics. It converts a signal from its original domain, often time or space, into a representation in the frequency domain (see Eq. (1)). The Fourier Transform decomposes a function (signal) into its constituent frequencies. It is based on the principle that any signal can be represented as a sum of simple sinusoids (sine and cosine functions) with varying amplitudes and frequencies. FT could be categorized into (i) continuous, which is used for continuous signals, it transforms a time-domain signal into a continuous spectrum of frequencies, and ii) discrete, which applies to discrete signals and is commonly used in digital signal processing. It computes the spectrum of a finite set of samples. How to perform Fourier Transform analysis using JS is provided below in Eq. (1) and **Box 1**.

```
// Step 1: Import the FFT library.
// This line imports the FFT (Fast Fourier Transform) library 'fft.js'.
// The FFT algorithm is used for computing the Discrete Fourier Transform (DFT) and its inverse.
const FFT = require('fft.js');
// Step 2: Define an example function to transform
// This function 'exampleFunction' is a sample function for which we want to perform the Fourier //
Transform.
// Here, it's defined as a sine function. Replace this with any function you want to analyze.
function exampleFunction(t) {
return Math.sin(t);}
// Step 3: Set the number of samples
// 'n' represents the number of samples. For FFT, it's best to use a power of 2 for efficiency.
// Here, 1024 samples are used.
const n = 1024;ural logarithm, i
// Step 4: Create an FFT instance
// This line creates a new FFT instance 'f' with 'n' points.
const f = new FFT(n);
// Step 5: Sample the function and prepare the input array
// This step involves creating an array of 'n' samples by evaluating the 'exampleFunction'
// at equally spaced intervals over 2π. This array 'input' represents the signal in the time domain.const
input = new Array(n).fill(0).map((_, idx) => exampleFunction(2 * Math.PI * idx / n));
// Step 6: Perform the FFT.Here, the FFT is performed on the input data.
// 'output' is an array that will store the FFT result. It's initialized using 'createComplexArray'.
'realTransform' performs the actual FFT computation on the 'input' and stores the result in 'output'.
'completeSpectrum' computes the full frequency spectrum.
const output = f.createComplexArray();
f.realTransform(output, input);
```

```
f.completeSpectrum(output);
// The 'output' array now contains the result of the FT and represents the frequency //domain
representation of the original time-domain signal.
```

Box 1.
The code that explains the FT steps needed for conducting it.

$$F(\omega) = \int f(t)\, e^{(-i\omega t)} dt \tag{1}$$

where, $F(\omega)$ is the Fourier Transform of $f(t)$. $\int$ denotes the integral over the entire domain of $f(t)$, which is usually time (t). $f(t)$ is the original function (or signal) being transformed. $e^{(-i\omega t)}$ is the complex exponential function, where e is the base of the imaginary unit, ω (omega) is the angular frequency.

4.2.2 Z-Transforms

The Z-Transform is a mathematical technique used in digital signal processing and control systems to convert discrete-time signals into a complex frequency domain, simplifying the analysis and design of digital systems, particularly for stability and frequency response.

4.2.3 Laplace Transforms

Laplace Transforms is a mathematical technique used to transform complex, time-based functions into a simpler, frequency-based domain [28]. This method is particularly valuable in engineering and physics for analyzing linear time-invariant systems. By converting differential equations, which often describe dynamic systems, into algebraic equations in the Laplace domain, simplifies the process of solving these equations. The Laplace Transform provides a powerful tool for understanding and manipulating the behavior of systems over time.

4.2.4 Cosine and Sine

Cosine and Sine Transforms, especially the Discrete Cosine Transform (DCT), are crucial for image compression and analysis, prominently used in JPEG compression [19]. DCT converts image data from the spatial to the frequency domain, focusing on lower frequency components vital for human vision, enabling efficient compression by preserving important visual information and discarding less perceptible details. While the Discrete Sine Transform (DST) is less commonly used than DCT, both play key roles in reducing image data redundancy and correlations, essential for compression, feature extraction, and noise reduction. These transforms are also vital for filtering, edge detection, and pattern recognition in digital image processing, improving the processing and analysis of digital images.

4.2.5 Hilbert Transforms

The Hilbert Transform is a crucial tool in digital image processing for converting real signals into complex ones, encapsulating amplitude and phase details. This is particularly useful for enhancing edge detection and representation in images, facilitating the analysis of textures and patterns through the extraction of instantaneous amplitude and phase information.

4.2.6 Frequency filtering

Frequency filtering is a technique for enhancing or suppressing specific features of an image by manipulating its frequency components [15]. Low-pass filters are used for smoothing and noise reduction by retaining low-frequency components, while high-pass filters are employed for edge enhancement and sharpening by preserving high frequencies. This method is critical for tasks such as noise reduction, edge detection, image sharpening, and feature extraction, playing a key role in both the preprocessing and post-processing stages of image analysis.

5. Image transformation and analysis

Image transformation and analysis encompass a range of operations aimed at extracting meaningful information from images. These transformations may include (i) image segmentation divides the image into meaningful segments or regions, (ii) feature extraction focuses on identifying and extracting specific features or patterns from the image, (iii) object recognition entails categorizing objects or patterns in the image, (iv) classification assigns pixels or regions to predefined classes or categories and (v) morphological operations analyze and process image structures based on shape and form. These operations collectively contribute to a comprehensive approach for understanding and interpreting the content of digital images, facilitating tasks such as object identification and image classification.

5.1 Satellite image classification

Image classification involves sorting and labeling pixels into predefined categories [15]. In principle, it is possible to group similar pixels into informational classes that are of interest to the users. The pixels carry the spectral information which is normally captured as digital numbers based on the reflections or the emissions from different bands. The classified pixels represent the informational classes that are of interest to the users. It is possible to categorize image classification methods into three types: (i) supervised, (ii) unsupervised, and (iii) hybrid [18].

5.1.1 Supervised classification

It involves the use of known, labeled data to classify new, unlabeled data. It is a type of machine learning, in which the algorithm is "trained" on a dataset where the categories or outcomes are already known. The process involves selecting representative samples for each class (training data) and using these to guide the algorithm in recognizing similar patterns in the rest of the data [27]. Supervised classification is notably effective for precise, well-labeled training datasets. This technique is widely implemented across various fields. However, our discourse will predominantly concentrate on the utilization of this technique for the delineation of land cover categories in satellite-based imagery. For the purpose of evaluating the precision of the model, the dataset is partitioned into training and testing subsets. For the purpose of verifying and evaluating the precision of the model, the dataset is partitioned into training and testing subsets. The training set educates the model to recognize patterns, while the testing set, unseen during training, is used to assess model performance, ensuring an unbiased evaluation of its practical applicability.

5.1.2 Unsupervised classification

On the other hand, the unsupervised classification does not require any prior labeling of data. Instead, the algorithm independently analyses the data to find patterns or clusters based on inherent similarities and differences. This approach is applied when predefined categories are absent, or during data examination to discover inherent groupings [27]. While processing can be relatively fast, the task of labeling the classes is challenging and time-intensive [23]. Common applications include identifying clusters in data for market research or ecological studies. The post-processing phase of unsupervised classification involves extensive techniques such as sieving, clumping, merging, and assigning significant categories to clusters. Additionally, this phase includes integrating the classified data with other datasets. One of the major challenges in this process is conducting accuracy assessments, due to the inherent nature of unsupervised classification.

5.1.3 Hybrid classification

It combines elements of both supervised and unsupervised methods. It may start with an unsupervised approach to explore and segment the data into clusters and then apply supervised techniques to label these clusters based on known samples.

In complex situations where neither solely supervised nor purely unsupervised approaches are adequate, this method can prove to be especially beneficial. Hybrid classification leverages the strengths of both approaches, providing a more nuanced and accurate classification in diverse datasets.

5.2 Accuracy assessment

Accuracy assessment evaluates the precision of a classification model or algorithm in predicting known outcomes, crucial for determining the model's reliability. In supervised classification, it assesses how well the model learns from training data and its accuracy on new data, using metrics such as overall accuracy and confusion matrix. An overall accuracy of over 80% is generally acceptable, but higher thresholds may be required for critical applications such as urban planning. These metrics offer a comprehensive view of model performance, particularly important in datasets with imbalanced class distribution.

6. Spatial cloud computing and GEE: harnessing the power of computational digital image processing

Spatial cloud computing is the type of cloud computing that is specifically designed to process and visualize spatial data [9]. It combines geospatial science and cloud computing in a way that leverages the power of distributed computing to enable geospatial discoveries [29]. There are three cloud computing service models: (i) Infrastructure as a Service (IaaS); (ii) Platform as a Service (PaaS); and (iii) Software as a Service (SaaS), which can be used for a variety of cloud computing applications, including spatial computing applications [30].

Google Earth Engine (GEE), serves as an exemplary case of spatial cloud computing, empowering users to access, explore, analyze, and visualize satellite imagery and remote sensing data hosted in the cloud. Google Earth Engine (GEE) is a unique PaaS, that uses Google's massive data catalog to analyze environmental and global data [31]. Google Earth Engine is used by scientists, researchers, and developers to study changes, trends, and variability on the Earth's surface [32]. GEE is considered by many researchers and users as a powerful tool in environmental monitoring and analysis, which may include climate change, biodiversity, land use and land cover changes, and disaster and risk management. GEE's power lies in its accessibility, cost-effectiveness, global reach, and cloud-based computing capabilities, making it an indispensable tool for environmental research and analysis.

In this chapter, we will delve into the utilization of GEE and JavaScript programming, using the code editor provided by GEE, for digital image processing and analysis of remotely sensed data.

6.1 Google earth engine common datasets

Google Earth Engine provides a constantly growing database of Earth observation images and scientific data, which is measured in petabytes. The public data archive includes more than forty years of historical images and scientific data (Google Earth Engine, 2023). The datasets provided by GEE catalog could be categorized into three themes: (1) Earth observation imagery, (2) atmospheric and weather, and (3) socio-economic data. For more in-depth information on this topic, please visit the Earth engine datasets catalog at:

https://developers.google.com/earth-engine/datasets/catalog

6.2 JavaScript and code editor: a gateway to Earth engine's geospatial computation power

To access Google Earth Engine through the code editor, you will need a Google account, a web browser with JavaScript enabled, and familiarity with JavaScript. For the most seamless and secure Google Earth Engine experience, we recommend Chrome. Its exceptional performance, compatibility, and security features make it the ideal choice for GEE users. To utilize the Google Earth Engine (GEE) code editor, a user must have a Google account and be registered with GEE, often receiving email confirmation within a few days. An uninterrupted internet connection and a contemporary web browser such as Chrome, are essential for accessing this online application. Chrome is recommended due to its superior optimization of web-based applications and integration with Google's services. A user's proficiency with the platform will be significantly enhanced by having foundational knowledge in JavaScript or Python and familiarity with GIS concepts. Once registered, the user can start exploring the various features of the code editor through its website. Visit code.earthengine.google.com, sign in with your Google account, and utilize the Earth Engine JavaScript API within the web-based IDE. The code editor offers features such as a map display, API reference documentation, and a git-based script manager for developing complex geospatial workflows.

7. Post-processing and visualization

Post-processing and visualization in digital image processing involves a set of operations performed on processed images to refine results and facilitate interpretation. These operations typically include (i) image fusion to combine information from multiple sources, (ii) image filtering to enhance or smooth specific features, (iii) rendering to visualize the final processed image, and (iv) quality assessment to evaluate the accuracy and reliability of the results [15, 16]. Post-processing is crucial for refining the output, ensuring optimal visual representation, and preparing the image for effective interpretation and analysis. The visualization step allows users to perceive and understand the processed data, providing insights into the underlying patterns or information present in the image.

8. Case study 1: satellite digital image analysis and data extraction using GEE and JS

8.1 Script overview

The methodology explained in the script (**Box 2**) is an intricate application of remote sensing techniques, implemented through the Google Earth Engine (GEE) and underscored by the use of band maths concepts. It focuses on the analysis of satellite imagery from the Landsat 9 series, employing band maths to develop indices such as the Normalized Difference Vegetation Index (NDVI), Water Index (WI), and Urban Index (UI). These indices are pivotal for the identification and delineation of diverse land cover features such as urban areas, vegetation, and water bodies within a specified region of interest (ROI), in this case part of Khartoum State. Leveraging the analytical power of GEE, this approach is instrumental for environmental monitoring and geographic analysis, providing nuanced insights into land use patterns and the effective management of natural resources. The process effectively showcases the utility of remote sensing and Geographic Information System (GIS) technologies in environmental and urban analysis. Through the strategic manipulation of different spectral bands, the script facilitates a comprehensive portrayal of the land cover characteristics within the chosen ROI. This level of analysis is invaluable across various domains, including urban planning, environmental conservation, agricultural monitoring, and natural resource management. The capacity to identify and categorize different land characteristics enhances our knowledge and strengthens the decision-making for future land utilization and conservation initiatives.

```
// Step 1: Define the Region of Interest (ROI)
// This defines the geographic area for the analysis, using coordinates to create a rectangle.
var roi = ee.Geometry.Rectangle([32.25, 15.25, 33, 16]);
// Step 2: Select Image Collection
// This selects the Landsat 9 image collection and filters it to include only images that intersect with the
ROI.
var collection = ee.ImageCollection('LANDSAT/LC09/C02/T1_L2')
.filterBounds(roi);
// Step 3: Filter by Date and Cloud Cover, and Select the Image with Least Cloud Cover
// This filters the image collection to a specific date range and sorts by cloud cover, selecting the image
with the least cloud cover.
collection = collection.filterDate('2022-01-01', '2023-12-31');
```

```
collection = collection.sort('CLOUD_COVER');
var leastCloudyImage = collection.first();
// Apply scaling factors function
// This function scales optical and thermal bands to enhance image visualization.
function applyScaleFactors(image) {
var opticalBands = image.select('SR_B.').multiply(0.0000275).add(-0.2);
var thermalBands = image.select('ST_B.*').multiply(0.00341802).add(149.0);
return image.addBands(opticalBands, null, true)
        .addBands(thermalBands, null, true);}
var scaledImage = applyScaleFactors(leastCloudyImage);
// Step 4: Clip, Create True Color Composite, and Add to Map
// This clips the image to the ROI and creates a True Color Composite for visualization.
var clippedImage = scaledImage.clip(roi);
var visualization = {bands: ['SR_B4', 'SR_B3', 'SR_B2'], min: 0.0, max: 0.3};
var trueColorComposite = clippedImage.visualize(visualization);
Map.addLayer(trueColorComposite, {}, 'TrueColor Composite');
// Step 5: Create and Add False Color Composite to Map
// This creates a False Color Composite using specific bands to highlight different features like vegetation
and water.
var falseColorComposite = clippedImage.visualize({bands: ['SR_B5', 'SR_B4', 'SR_B3'], min: 0.0, max:
0.3});
Map.addLayer(falseColorComposite, {}, 'FalseColor Composite');
// Step 6: Urban Area Identification
// This identifies urban areas using a spectral index and masks non-urban areas for //visualization.
var urbanAreas = clippedImage.normalizedDifference(['SR_B4', 'SR_B3']).gt(0.2);
var urbanMask = urbanAreas.updateMask(urbanAreas);
Map.addLayer(urbanMask, {palette: ['red']}, 'Urban Areas');
// Step 7: Vegetation Delineation
// This delineates vegetation using the NDVI index and masks non-vegetation areas for visualization.
var ndvi = clippedImage.normalizedDifference(['SR_B5', 'SR_B4']).gt(0.3);
var vegetationMask = ndvi.updateMask(ndvi);
Map.addLayer(vegetationMask, {palette: ['green']}, 'Vegetation Areas');
// Step 8: Water Body Delineation
// This delineates water bodies using the NDWI index and masks non-water areas for visualization.
var ndwi = clippedImage.normalizedDifference(['SR_B3', 'SR_B5']).gt(0.3);
var waterMask = ndwi.updateMask(ndwi);
Map.addLayer(waterMask, {palette: ['blue']}, 'Water Bodies');
// Center the map on the ROI
Map.centerObject(roi);
```

Box 2.
Explaining the coding steps in case study.

The script presented in **Box 2** outlines eight actionable steps that users can directly implement in the GEE (Google Earth Engine) code editor. This is designed to engage users in understanding the significance of image processing through GEE and its JavaScript API. Moreover, the user is equipped to derive essential insights from the imagery by applying band maths methods. This process includes the use of prominent spectral indices, such as the Normalized Difference Vegetation Index (NDVI), Aquatic Spectral Indices for water analysis, and Anthropogenic Land Use Indices for urban area delineation. These indices play a critical role in the nuanced assessment and interpretation of satellite imagery data. **Box 2** explains the coding steps in case study 1. The code could be accessed and reused from https://github.com/Elmahal/DIP_GEE_JS_Codes

8.2 Case study 1: model outputs

In our first case study, we delve into extracting data from Landsat 9 images from 2022, focusing on processing techniques and region of interest clipping (ROI), as detailed in our code snippet. This includes generating true and false color composites (**Figure 1**). The aim is to identify, visualize, and extract information for essential environmental elements such as vegetation and water bodies using normalized difference indices (NDVI and NDWI, respectively), displayed in green and blue (**Figure 2**). Improving the quality of the extracted information could include optimizing spectral band selection to match the study area's unique characteristics, adjusting index thresholds for clearer feature differentiation, and integrating more data layers or indices for comprehensive environmental analysis.

9. Case study 2: advanced LULC analysis through supervised classification of Sentinel-2 imagery

The Sentinels, satellite constellations crafted by the European Space Agency (ESA), incorporate both radar and optical imaging capabilities, making them well-suited for tasks such as environmental surveillance, climate monitoring, and air quality assessment [33]. In this chapter, we will explore together the data provided by Sentinel-2, the multispectral instrument (MSI). The MSI offers high-resolution multispectral imagery covering various wavelengths in the electromagnetic spectrum. The data are crucial for monitoring vegetation, soil, and water resources, as well as

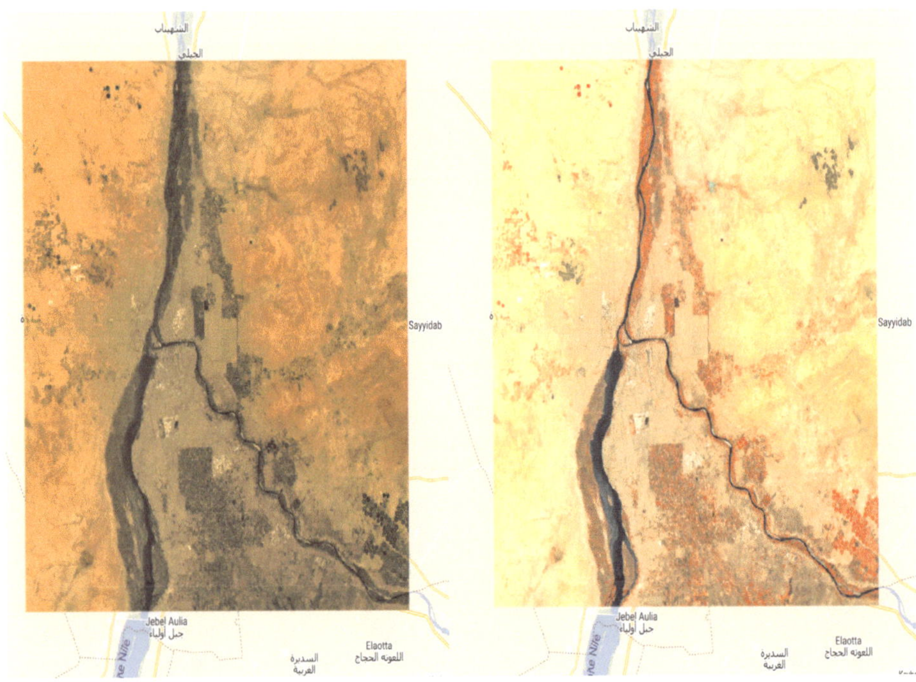

Figure 1.
True color composite (left) and false color composite (right).

DOI: http://dx.doi.org/10.5772/intechopen.1004501

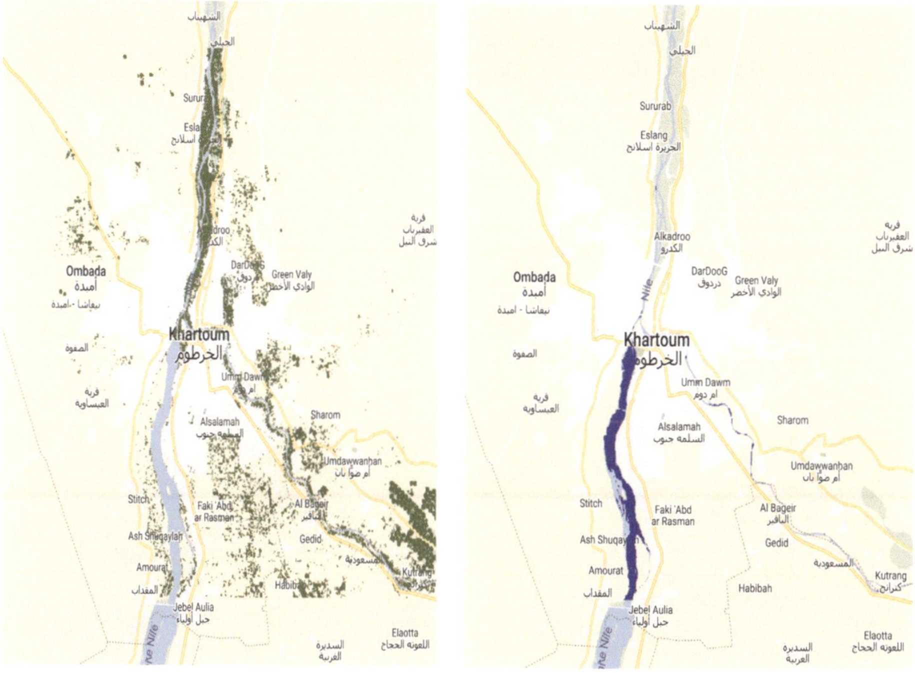

Figure 2.
Information extraction for vegetation (left) and water (right) using NDVI and NDWI, respectively.

detecting and analyzing changes in land cover (The Sentinel missions, 2023) [33]. The information provided by Sentinel-2 supports informed decision-making in environmental management, conservation, and disaster response, making it a valuable tool for humanitarian efforts and mitigating the impact of natural disasters (**Box 3**).

```
// Importing the region of interest (roi) – Greater Khartoum
var roi = ee.FeatureCollection('projects/ee-anwareltayeb2/assets/KRT_SA'); // //"KRT_SA", is the region
of our interest but could be replace by your own roi
// Import the five classes for classification
exports.water = water;
exports.agric = agric;
exports.ind = ind;
exports.urban = urban;
exports.barren = barren;
// Load Sentinel-2 ImageCollection for the specified date range and region
var image = ee.ImageCollection('COPERNICUS/S2_SR_HARMONIZED')
        .filterDate('2023-01-01', '2023-04-15')
        .filterBounds(KRT_SA)
        .select(['B4', 'B3', 'B2', 'B8']) // Selecting specific bands
        .median();
// Visualisation parameters
var visParamsFalse = {
bands: ['B4', 'B3', 'B2'],
min: 0.0,
max: 3000,
gamma: 1.4
};
```

```
// Add image layer to the map for visualization
Map.addLayer(image.clip(KRT_SA), visParamsFalse, 'KRTM23');
Map.centerObject(KRT_SA, 10);
// Supervised classification with CART
// Define labels and bands
var label = 'Class';
var bands = ['B2', 'B3', 'B4', 'B8'];
// Assign class labels to each feature in the collections
var waterWithClass = water.map(function(feature) {
return feature.set('Class', 0);
});
var agricWithClass = agric.map(function(feature) {
return feature.set('Class', 1);
});
var indWithClass = ind.map(function(feature) {
return feature.set('Class', 2);
});
var urbanWithClass = urban.map(function(feature) {
return feature.set('Class', 3);
});
var barrenWithClass = barren.map(function(feature) {
return feature.set('Class', 4);
});
// Merge the class collections into one dataset for training
var training = waterWithClass.merge(agricWithClass)
.merge(indWithClass)
.merge(urbanWithClass)
.merge(barrenWithClass);
print(training); // Check the training dataset
// Prepare the input image for classification
var input = image.select(bands);
// Sample the regions for training the classifier
var trainImage = input.sampleRegions({
collection: training,
properties: [label],
scale: 10
});
print(trainImage);
// Split the data into training and testing sets
var trainingData = trainImage.randomColumn();
var trainSet = trainingData.filter(ee.Filter.lessThan('random', 0.8));
var testSet = trainingData.filter(ee.Filter.greaterThanOrEquals('random', 0.8));
// Train the classifier
var classifier = ee.Classifier.smileCart().train(trainSet, label, bands);
// Classify the image
var classified = input.classify(classifier);
print(classifier.getInfo());
// Define a palette for the land cover classes
var landcoverPalette = [
"#0000FF", // water
"#00FF00", // agriculture
"#800080", // industrial
"#FFFF00", // urban
"#D2B48C" // barren
];
// Add classified layer to the map
Map.addLayer(classified.clip(KRT_SA), {palette: landcoverPalette, min: 0, max: 5}, "classified");
// Accuracy assessment and confusion matrix
```

```
var confusionMatrix = ee.ConfusionMatrix(testSet.classify(classifier)
.errorMatrix({
actual: 'Class',
predicted: 'classification'
}));
print("ConfusionMatrix:", confusionMatrix);
print("Overall Accuracy:", confusionMatrix.accuracy());
// Calculating areas of the created classes in GEE
var areaImage = ee.Image.pixelArea().addBands(classified);
var areas = areaImage.reduceRegion({
reducer: ee.Reducer.sum().group({
groupField: 1,
groupName: 'class',
}),
geometry: roi.geometry(),
scale: 500,
maxPixels: 1e10
});
print(areas);
// Processing the area calculations
var classAreas = ee.List(areas.get('groups'));
var classAreaLists = classAreas.map(function(item) {
var areaDict = ee.Dictionary(item);
var classNumber = ee.Number(areaDict.get('class')).format();
var area = ee.Number(areaDict.get('sum')).divide(1e6).round();
return ee.List([classNumber, area]);
});
var result = ee.Dictionary(classAreaLists.flatten());
print(result);
```

Box 3.
The comprehensive code for case study 2 with elaborated comments.

9.1 Get your hands wet with LULC supervised classification

In this section, the user will be guided through the detailed process of utilizing the Google Earth Engine code editor, JavaScript programming, and Sentinel-2 imagery. The emphasis will be on the application of supervised image classification techniques for LULC analysis utilizing machine learning algorithms. The chosen study area for this application is part of Khartoum state in Sudan, a country currently experiencing profound human challenges. This region's selection underscores both its geographic significance and the current humanitarian context. The goal is to classify the LULC into distinct categories such as urban areas, agricultural land, forests, and water bodies.

A vital step in the classification process is the overall accuracy assessment of the model. This involves evaluating how well the model performs in correctly classifying various LULC types, ensuring the accuracy of the chosen algorithm and the exactness of the classification results.

This is achieved through the confusion matrix, which rigorously evaluates the model's performance. The overall accuracy of 80% is considered good for most applications [34]. However, many factors can play a role in the accuracy of the model's performance. This section highlights the effective use of spatial cloud computing in the detailed examination and classification of various land types and their utilizations within the selected region of interest. The final outcome is the development of a comprehensive land use and land cover (LULC) map, achieved through the application of Google Earth Engine (GEE) for both visualization and quantitative analysis.

9.2 Accessing Sentinel data *via* GEE/JS API

Unlike traditional approaches to manipulating Earth Observation (EO) data, GEE provides novel methods to handle voluminous data, with a variety of formats, processing, and analyzing it without having a developed computer system [11]. To access the code editor and interact with the GEE platform, JavaScript serves as one of the native GEE APIs. A simple way of accessing the Sentinel data is to rely on the available codes provided by the Earth Engine Data Catalog (https://developers.google.com/earth-engine/datasets). Once the user accesses the code *via* the code editor, it is possible to customize it and build on it according to the user's need. But before that the user needs to understand the components of the Sentinel data. The Sentinel-2 MSI data consists of (i) bands, (ii) quality indicators, and (iii) metadata (https://sentinels.copernicus.eu/web/sentinel/user-guides/sentinel-2-msi).

9.3 Understanding the LULC supervised classification step by step

In the following paragraphs, we are going to explain the codes used for supervised classification of satellite imagery from Sentinel-2, specifically for a study area (ROI) in Greater Khartoum, Sudan. It includes image preparation, classification, accuracy assessment, and area calculation for different LULC classes. We will divide the process into individual steps, making it easier for the user to follow each part one at a time.

9.4 Loading Sentinel-2 image collection

The process begins by loading a Sentinel-2 image collection, where images are filtered by date and specific bands (10-meter resolution) are selected. This step is crucial for capturing high-resolution, relevant spectral data. The median of these images is computed to reduce noise and cloud cover, ensuring a clear and consistent dataset for analysis.

9.5 Selecting the region of interest (ROI)

Selecting the ROI involves balancing ideal study requirements against practical data limitations. GIS tools are instrumental in defining and visualizing ROIs. In this case, the ROI "KRT_SA" is predefined and imported as a feature collection, representing the study area.

9.6 Defining visualization parameters

Visualization parameters are set using the "visParamsFalse" object. This configuration dictates how the Google Earth Engine represents satellite imagery on the map. It utilizes bands 4, 3, and 2 (red, green, blue) for true-color representation. The pixel value range and gamma adjustment enhance contrast and detail, optimizing the visual interpretation of the image.

9.7 Adding the visualized layer to the basemap

The processed image is added to the map as a layer with custom visualization settings and labeled for easy identification. The map is then centered on the ROI at a defined zoom level, focusing on the display on the area of interest.

9.8 Preparing training areas for classification

For this case study, five land use and land cover (LULC) classes are identified, each with a unique hexadecimal color code. Training areas could be imported or digitized on the map, and each is assigned a class label corresponding to a specific land cover type.

9.9 Key factors in choosing training areas

Selecting training areas requires consideration of various factors to ensure accurate classification. These include capturing the full spectrum of each class, ensuring geographic distribution, maintaining temporal consistency, verifying the accuracy of training data, understanding spectral properties, and considering landscape dynamics.

9.10 Data preparation for supervised classification

Data for supervised classification is prepared by labeling different datasets according to their land cover type. This labeled data trains a classification model (CART algorithm) to predict land cover types based on spectral characteristics.

9.11 Merging classes

A comprehensive training dataset is prepared by merging datasets representing different classes into one large dataset. This method ensures a varied and representative sample for the machine learning model.

9.12 Implementing supervised classification with CART algorithm

The script selects relevant bands, samples regions based on labeled training data, randomizes these samples, and splits them into training and testing sets. A CART classifier is then trained, typical in remote sensing applications for tasks such as land cover classification.

9.13 Visualizing land cover classification results

The script classifies an input image using a trained classifier, assigns a color palette for different land cover classes, and adds the classified image to the map. This visualization aids in interpreting satellite imagery.

9.14 Evaluating the classification model performance

Model accuracy is assessed using a confusion matrix, which evaluates performance on a test dataset. This step is vital in understanding and validating the classification model's effectiveness.

9.15 Class-based area analysis in satellite imagery using Google Earth Engine

The concluding phase entails the examination and classification of specific zones within satellite imagery into distinct categories, each representing a different type of land cover. This process exemplifies the conversion of unprocessed satellite images into significant, systematically categorized spatial data for use in environmental and geographic research. The code presented is an advanced instance of spatial image processing through Google Earth Engine (GEE). It targets a designated area of interest (ROI) in the image and utilizes the "reduceRegion" function to aggregate pixel values in this region. A critical element of this analysis is the aggregation of pixel values into designated classes, representing various land cover types, vegetation indices, or other ecological attributes. This analytical method is refined through settings such as spatial resolution (scale) and maximum pixel limits. It further manipulates this list to translate the area calculations into a more understandable unit, such as square kilometers, correlating them with their corresponding class labels.

Ultimately, the script generates a meticulously arranged dictionary, associating each class with its respective area size. This output is instrumental for tasks in environmental surveillance, resource management, and land utilization studies. The code showcases the robust capability of GEE in converting basic satellite images into informative, classified spatial data crucial for ecological and geographical research. The code could be accessed and reused from https://github.com/Elmahal/DIP_GEE_JS_Codes

9.16 Case study 2: model outputs

The focus of case study 2 is to explore the output and the performance of a supervised classification model, which is CART algorithm. The classifier is

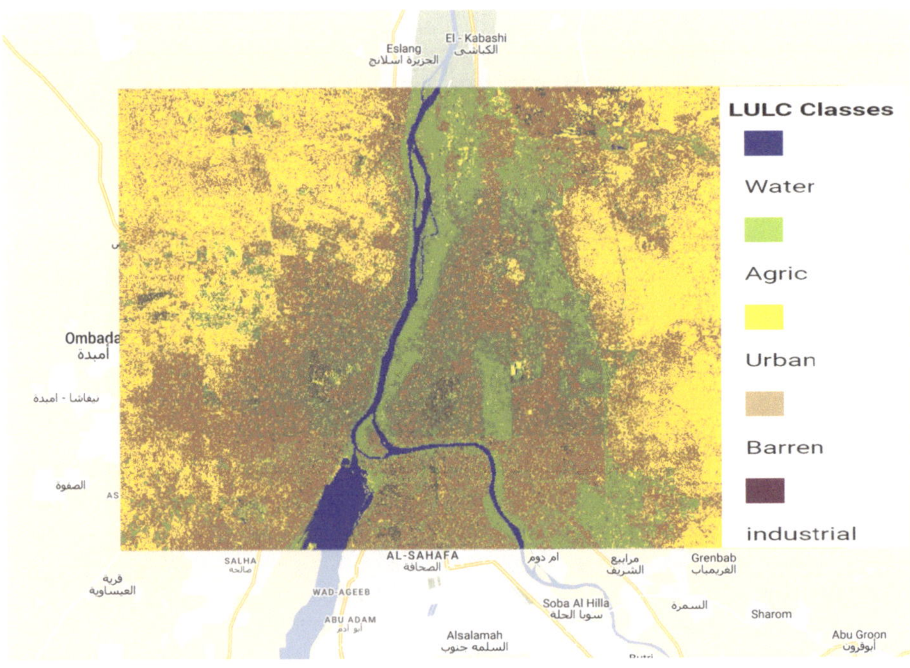

Figure 3.
The output of the LULC classes overlaid over Google map.

Actual	Correct predictions	Total actual	Accuracy (%)
Water	90	100	90.0
Agriculture	80	100	80.0
Industrial	85	100	85.0
Urban	80	100	80.0
Barren	95	100	95.0

Table 1.
Per class accuracy for evaluating the CART model performance.

trained with labeled data [water, agriculture, industrial, urban, and barren lands]. The result of the output is shown in **Figure 3**. Furthermore, the classification's effectiveness is assessed *via* a confusion matrix, which evaluates the overall accuracy which was found to be 85% a globally accepted accuracy in LULC classification. The accuracy per class is displayed in **Table 1**. The accuracy assessment is a requirement in any classification process and it is crucial for understanding the classifier's performance in accurately identifying each land cover type. Improving the model's performance can be achieved through several strategies. One approach is to enhance both the quantity and quality of the training data. Another method involves incorporating a wider variety of representative samples from each class to ensure a comprehensive learning base. Additionally, expanding the selection of spectral bands used in the analysis can provide more detailed information for the model to process. Lastly, integrating additional features such as elevation and texture into the model can significantly improve its ability to classify data accurately.

10. Conclusion

In this chapter, we have delved into the nuances of digital image processing by leveraging the capabilities of JavaScript, Google Earth Engine (GEE), and the code editor. This exploration bridges the gap between theoretical concepts and their practical application in analyzing and visualizing remotely sensed data. The chapter is enhanced by including two case studies, offering practical experience to the readers. These studies act as basic exercises, accommodating the needs of both novice and experienced users. They demonstrate the practical applications of digital image analysis concepts, with JavaScript playing a key role in showcasing its adaptability and efficacy in processing remote sensing data. The skills and insights gained from this chapter are not limited to JavaScript; they are applicable across various programming languages, making them versatile for a broad audience. This is particularly relevant for users of Python or R, who can adapt these techniques to their preferred programming environments. The chapter emphasizes the effective use of JavaScript API and Google Earth Engine in remote sensing and digital image analysis, setting a foundation for future advancements in geospatial data analysis across fields, such as research, environmental monitoring, and resource management. It encourages further exploration and learning in the evolving field of digital image analysis and its significant potential in remote sensing.

Author details

Anwarelsadat Elmahal* and Eltaib Ganwa
Faculty of Geography and Environmental Sciences, University of Khartoum, Khartoum, Sudan

*Address all correspondence to: anwareltayeb@gmail.com

References

[1] Feldman H. Sputnik: The First Satellite. New York: The Rosen Publishing Group, Inc; 2002

[2] Dickson P. Sputnik: The Shock of the Century. New York: U of Nebraska Press; 2019

[3] Jakhu RS, Pelton JN. Global Space Governance: An International Study. Berlin: Springer; 2017

[4] de Carvalho Alves M, Sanches L, Silva de Menezes F, Trindade LRSLC. Multisensor analysis for environmental targets identification in the region of Funil dam, state of Minas Gerais, Brazil. Applied Geomatics. 2023;**15**(4):807-827. DOI: 10.1007/s12518-023-00523-w

[5] Varshney PK, Arora MK. Advanced Image Processing Techniques for Remotely Sensed Hyperspectral Data. Berlin: Springer Science & Business Media; 2013

[6] Chen L-C, Papandreou G, Kokkinos I, Murphy K, Yuille AL. Semantic Image Segmentation with Deep Convolutional Nets and Fully Connected CRFs. 2016. DOI: 10.48550/arXiv.1412.7062

[7] Zhu XX, Tuia D, Mou L, Xia G-S, Zhang L, Xu F, et al. Deep learning in remote sensing: A comprehensive review and list of resources. IEEE Geoscience and Remote Sensing Magazine. 2017;**5**: 8-36. DOI: 10.1109/MGRS.2017.2762307

[8] Ma L, Liu Y, Zhang X, Ye Y, Yin G, Johnson BA. Deep learning in remote sensing applications: A meta-analysis and review. ISPRS Journal of Photogrammetry and Remote Sensing. 2019;**152**:166-177. DOI: 10.1016/j.isprsjprs.2019.04.015

[9] Gorelick N, Hancher M, Dixon M, Ilyushchenko S, Thau D, Moore R. Google earth engine: Planetary-scale geospatial analysis for everyone. Remote Sensing of Environment, Big Remotely Sensed Data: Tools, Applications and Experiences. 2017;**202**:18-27. DOI: 10.1016/j.rse.2017.06.031

[10] Yan J, Ma Y, Wang L, Choo K-KR, Jie W. A cloud-based remote sensing data production system. Future Generation Computer Systems. 2018;**86**: 1154-1166. DOI: 10.1016/j.future.2017.02.044

[11] Elmahal AE, Musa MMI, Elmahal AE, Musa MMI. Spatial Cloud Computing Using Google Earth Engine and R Packages. London, UK: IntechOpen; 2023. DOI: 10.5772/intechopen.1002686

[12] Weinman J. Cloudonomics: The Business Value of Cloud Computing. New Jersey: John Wiley & Sons; 2012

[13] Davison D. The Total Economic Impact of Google Cloud Platform. Forrester Research; 2020

[14] Jensen J. Remote Sensing of the Environment: An Earth Resource Perspective. 2nd ed. Upper Saddle River, NJ: Pearson; 2006

[15] Campbell JB, Wynne RH. Introduction to Remote Sensing. 5th ed. New York: Guilford Press; 2011

[16] Lillesand T, Kiefer RW, Chipman J. Remote Sensing and Image Interpretation. New Jersey: John Wiley & Sons; 2015

[17] Curlander JC, McDonough RN. Synthetic Aperture Radar: Systems and Signal Processing. New Jersey: Wiley; 1992

[18] Borra S, Thanki R, Dey N. Satellite Image Analysis: Clustering and Classification. Berlin: Springer; 2019

[19] Manolakis DG, Lockwood RB, Cooley TW. Hyperspectral Imaging Remote Sensing: Physics, Sensors, and Algorithms. Cambridge: Cambridge University Press; 2016

[20] Favorskaya MN, Jain LC. Handbook on Advances in Remote Sensing and Geographic Information Systems: Paradigms and Applications in Forest Landscape Modeling. Berlin: Springer; 2017

[21] Al-Fares W. Historical Land Use/ Land Cover Classification Using Remote Sensing: A Case Study of the Euphrates River Basin in Syria. Berlin: Springer Science & Business Media; 2013

[22] Moigne JL, Netanyahu NS, Eastman RD. Image Registration for Remote Sensing. Berlin: Cambridge University Press; 2011

[23] Aronoff S. Remote Sensing for GIS Managers. Redlands, California: ESRI Press; 2005

[24] Alves M d C, Sanches L. Remote Sensing and Digital Image Processing with R. Boca Raton, Florida: CRC Press; 2023

[25] Soergel U. Review of radar remote sensing on urban areas. In: Soergel U, editor. Radar Remote Sensing of Urban Areas, Remote Sensing and Digital Image Processing. Netherlands, Dordrecht: Springer; 2010. pp. 1-47. DOI: 10.1007/ 978-90-481-3751-0_1

[26] Spiteri A. Remote sensing 96: Integrated applications for risk assessment and disaster prevention for the mediterranean. In: Proceedings of the 16th EARSeL Symposium, Malta, 20–23 May 1996. Boca Raton, Florida: CRC Press; 1997

[27] Varshney PK, Arora MK. Advanced Image Processing Techniques for Remotely Sensed Hyperspectral Data. Berlin: Springer Science & Business Media; 2004

[28] Kuhfittig PKF. Introduction to the Laplace Transform. Berlin: Springer Science & Business Media; 2013

[29] Yang C, Goodchild M, Huang Q, Nebert D, Raskin R, Xu Y, et al. Spatial cloud computing: How can the geospatial sciences use and help shape cloud computing? International Journal of Digital Earth. 2011;**4**:305-329. DOI: 10.1080/17538947.2011.587547

[30] Mell P, Grance T. The NIST Definition of Cloud Computing NIST (Special Publication (SP) 800-145). Gaithersburg, Maryland: National Institute of Standards and Technology; 2011. DOI: 10.6028/NIST.SP.800-145

[31] Aybar et al. Rgee: An R package for interacting with Google Earth Engine. Journal of Open Source Software. 2020; **5**(51):2272. DOI: 10.21105/joss.02272

[32] Google Earth Engine. Google Earth Engine [Online]. Available from: https:// earthengine.google.com/ [Accessed: January 30, 2023]

[33] MSI Instrument - Sentinel-2 MSI Technical Guide. Sentinel Online, European Space Agency [Online]. Available from: https://sentinels.copernicus.eu/web/sentinel/technical-guides/sentinel-2-msi/msi-instrument [Accessed: January 30, 2023]

[34] Olson DL, Delen D. Advanced Data Mining Techniques. Berlin: Springer Science & Business Media; 2008

Chapter 4

Use of Drones to Monitor and Aid Disaster Management Efforts of Natural Resources

Daniel R. Unger, David L. Kulhavy, I-Kuai Hung, Victoria Williams, Yanli Zhang and Reid Viegut

Abstract

Natural disasters, such as tornados, high wind damage, lightning strikes, ice storms, snow storms, and historic flood events, can cause extensive damage to natural resources. Drones, with their ability to observe the surface of the earth from a distance can provide individuals involved with disaster management evaluation and cleanup effort's information not accessible or not possible with a traditional field-based site visit. Faculty within the Arthur Temple College of Forestry and Agriculture at Stephen F. Austin State University are researching and demonstrating how drones can be used in a host of natural resource activities including assisting disaster management efforts.

Keywords: natural resource, disaster management, drone, storm, FEMA

1. Introduction

The use of Unmanned Aerial Systems (UAS) or drones, with their ability to observe the surface of the earth from a distance using either a nadir or oblique perspective, can provide individuals involved with disaster management evaluation and cleanup efforts information not attainable or not possible with a traditional field-based site visit. Drones can provide individuals involved in disaster management activities unique information only obtainable from an aerial perspective.

Faculty within the Arthur Temple College of Forestry and Agriculture (ATCFA) at Stephen F. Austin State University (SFASU) are researching and demonstrating how drones can be used in a host of natural resource activities including assisting disaster management efforts. Natural resource education for the 21st century must maintain rigor, relevance and building relationships [1]. Managing natural resources, which includes using geospatial technologies, relies heavily on the use of drones to map, monitor and quantify natural resources and their applicability to unexpected events such as natural disasters [2]. The use of drones in disaster management reduces time for locating victims; has a rapid response, provides quick evaluation, and employs needed sensor units for recovery [3].

Since drones are relatively new to the process of collecting data on natural resources, emphasis should be placed on training and preparing for this technology across all natural resource disciplines. "Society-ready" natural resource managers need to be capable of dealing effectively with all natural resources at their disposable and under their supervision. Natural resource graduates must be prepared to be able to effectively map, monitor, and quantify the natural resources under their management control. In addition, as technology and skill sets improve in the use of drone applications, teaching materials that focus on non-traditional use of drones such as disaster management activities should be integrated into natural resource curricula [4–6].

At SFASU there are four faculty members dedicated to the integration of drone technology within all natural resource endeavors. The four faculty members are identified as the SFASU "Drone Squad" and they focus their teaching, research, and service activities on using drones to effectively quantify and qualify forests and natural resources. In addition to traditional qualitative and quantitative assessment focus areas, they are heavily involved with researching the use of drones and how they can be used to increase the effectiveness of disaster management response activities.

The objective of this chapter is to present the reader how drones can be used to aid individuals involved with natural disaster management efforts. Examples provided demonstrate the effectiveness of using drones to help disaster management efforts within commonly occurring natural disasters including tornados, high wind damage, lightning strikes, ice storms, snow storms, and historic flood events.

2. Impact of tornado in Appleby Texas

A recent tornado that affected the small rural town of Appleby, Texas is a good example of how drones can be used to assess tornado damage from a distance using drones flown above tree top and below the maximum allowed flying height of 400 feet. Faculty from ATCFA, and student drone pilots, flew the tornado damaged area to assist the local rural community to visually assess the status of the disaster

Figure 1.
Drone image showing roof damage from a fallen tree.

Figure 2.
Drone image used to document path of tornado winds.

and reported the damage to the Federal Emergency Management Agency (FEMA) for damage assessment for Nacogdoches County.

Figure 1 demonstrates a nadir perspective from a drone, or straight down perspective, highlighting the visual perspective of a tree fallen on top of an extension of a house to assess roof damage unviewable from a ground view. **Figure 2** shows a drone image with a non-nadir view, also known as an oblique perspective, to visually demonstrate the visual analysis of not only a fallen tree caused by the tornado but also to document the direction of the wind.

Figure 3 demonstrates the use of a drone oblique image to visualize tornado tree damage amount in a larger geographic area other than a single home dwelling, while

Figure 3.
Drone image used to document tornado destruction near homes.

Figure 4.
Drone image demonstrating houses sparred from a downed tree.

Figure 4 shows how a tornado can damage individual trees within a natural resource ecosystem while simultaneously sparing housing structures.

3. Tree damage assessment in East Texas

Tree damage, from a variety of storms such as high winds and lightning strikes, can cause severe damage to individual trees which stand alone and are highly vulnerable out in the open. The campus of Stephen F. Austin State

Figure 5.
Drone image of tree damage high in the forest canopy.

Figure 6.
Drone image used to assess the effect of a tree lightning scar.

University, which is certified as a Tree USA Campus, has a large number of open trees that are vulnerable to storms.

After a recent storm, two members of the faculty "Drone Team" surveyed the campus for damage to the trees on campus. **Figure 5** demonstrated a drone image used to document the damage to the top of a large pine tree on campus as the results of an intense lightning strike. This demonstrates the ability of a drone to acquire horizontal images of a tree crown from the side high in the canopy for a perspective not available from ground level. The drone image displayed in **Figure 6** shows how drones can be flown high and in very close proximity to the bole of a tree to visual assess the effect of an intense lightning strike on the upper section of a tree stem. The damage was reported to the campus division of Emergency Management and the Facilities Management director and the tree was subsequently removed.

4. Tornado damage assessment in Cushing Texas

Demonstration of using drones to assess damage from another tornadic event, in this case within the rural community of Cushing, Texas, not only demonstrated the utility of drones to visually assess damage from a tornado but it also demonstrated the utility of drone imagery to create geometrically correct orthophoto mosaics that was used as base maps in disaster management documentation.

Figure 7 is a drone image of the aftermath of the Cushing, Texas tornado demonstrating an oblique view of severe tree damage adjacent to severe damage to a medium sized building. **Figure 8** demonstrates the use of drone imagery acquired high above the forest canopy to assess tornado damage within a portion of the affected community containing multiple housing structures. **Figure 9** shows a nadir view taken high

Figure 7.
Drone image of severe tree damage to a building.

Figure 8.
High altitude drone image showing area wide tornado damage.

above a fallen tree showing the destruction from the tree as it fell on a small house causing severe damage.

Figure 10 is an orthophoto mosaic of a section of the Cushing, Texas tornado path. An orthophoto mosaic is an image obtained by combining multiple nadir drone images into a composite image using geospatial science software to create a highly accurate and geographically correct image not attainable from a single drone image. Orthophoto mosaics are created from multiple individual drone images to not only

Figure 9.
Severe housing damage seen in a nadir drone image.

Figure 10.
Orthophoto mosaic of the Cushing, Texas tornado path.

create highly accurate visualization of a disaster management area but are highly effective in disaster management planning efforts. The imagery indicates to the natural resource manager both the direction of the tornado, but also susceptibility of both hardwood and pine species.

5. Caddo Mounds state historical site tornado damage assessment

In addition to damaging physical natural resources like a tree or a river ecosystem, natural disasters can affect cultural resources as well. Recently in East Texas a very high impact tornado west of Alto, Texas with a wind speed of 257.5 km/hour, destroyed part of the structures at Caddo Mounds State Historical Site that was dedicated to the history of the Caddo Indian Nation, one of the original inhabitants of East Texas.

Immediately after the tornado destroyed portions of the buildings in Caddo Mounds State Historical site, ATCFA drone faculty and student pilots flew drones to capture imagery of the affected area to document the tornado destruction as an aid to planning the park restoration efforts [7].

Figure 11 shows the utility of using drones to document the extremely severe damage to the park visitor center where individuals took shelter during the tornadic event. **Figure 12**, a high oblique, and low altitude drone image, portrays the visitor center from a different angle while also showing the location of the Caddo House (Koo-Hoot Kiwat, named by the Caddo Nation Native Americans) location in the background that was destroyed by the tornado [7].

In addition to still images acquired from a drone, drones can also be used to record videos of a disaster area to provide a more realistic perspective not viewable within a single drone image. Video 1 is a flyover of the Caddo Mounds State Historical Site demonstrating the efficiency of using a drone video to portray the stark reality of a tornado's impact on a cultural resource. Caddo Nation survivors of the tornado are termed Shahó and each member was given a three-dimensional model of the Caddo House created from the drone imagery. In addition, the imagery was used in reconstruction of the site that is scheduled to reopen in 2024 [8].

Figure 11.
Visitor center damage seen from a high oblique drone image.

Figure 12.
Caddo house location before tornado impact.

6. Tree lightning strike fire assessment on the SFASU campus

Lightning strikes can cause severe damage to trees, especially for open grown trees that typically exist in urban areas such as the campus of SFASU. After a recent storm containing a series of severe lightning strikes on the SFASU campus, a faculty member within the ATCFA Drone Squad documented the effect of lightning strikes on pine trees on the SFASU campus.

Figures 13 and **14** demonstrate different drone images of a pine tree on fire as the result of a lightning strike on campus. Both drone images represent still imagery depicting the fire generated from the lighting strikes from different angles and obtained near the tree stem high up in the forest canopy, a perspective not attainable from the ground view.

If mid-stem drone images do not provide a disaster assessment team the required information, the drone can be flown further up the tree stem to acquire drone imagery from different attitudes and different cardinal directions providing the viewer a new perspective of a lightning scare (**Figure 15**).

Videos of trees stems on fire as the results of lightning scars, as seen in Videos 2 and 3, not only provide a video documentation of the disaster but provide the visual observer of videography documentation that is very compelling emotionally when describing the effect of a tree stem fire.

7. Ice storm damage assessment in East Texas

Damage to natural resources, especially within a forest ecosystem, can be very traumatic due to heavy ice accumulation. Recently, East Texas experienced an intense

Figure 13.
Drone image of pine tree on fire; view 1.

Figure 14.
Drone image of pine tree on fire; view 2.

Figure 15.
Drone image of tree lightning scare high in forest canopy.

ice storm that caused catastrophic damage to the local forest ecosystem which was documented by various Drone Squad faculty members.

Figure 16 is a high-altitude landscape perspective of an East Texas forest ecosystem depicting the extent of a recent ice storm at the landscape level. **Figure 17** portrays a more

Figure 16.
Drone image depicting ice storm damage at the landscape level.

Figure 17.
Drone image showing bending of trees due to heavy ice accumulation.

localized effect of the ice storm as seen in a lower oblique drone image revealing the ice collected on the trees while causing severe bending of tree stems of younger pine trees.

Figure 18 shows an above ground level and below canopy level drone flown to document the damage to trees within the ecosystem by documenting the amount of trees stems that broke due to the heavy weight of the accumulating ice on tree branches. **Figure 19** is also an above ground and below canopy drone shot within a forest ecosystem to demonstrate the bending of young pine trees due to heavy ice accumulation.

Video 4 demonstrates the utility of using a drone to fly from ground level up to canopy level and further to document ice storm damage not only within a small geographic area but how it expands beyond a local area as well. Video 5 portrays a video obtained

Figure 18.
Drone image used to quantify tree branches lost in an ice storm.

Figure 19.
Drone image of young pine trees bent due to ice accumulation.

from a done flying low above forest canopy height to demonstrate the effect of ice storm damage using videography which can portray tree ice damage more emotionally.

8. Snow storm damage assessment in East Texas

Damage to natural resources can also be traumatic to a forest ecosystem due to heavy snow accumulation, similar to ice accumulation in the previous section. Recently in East Texas an intense snowstorm caused catastrophic damage to the local forest ecosystem which was documented by various Drone Squad faculty members.

Figure 20.
Drone image depicting snow storm damage at the landscape level.

Figure 20 is a high-altitude landscape perspective of an East Texas forest ecosystem depicting the extent of a recent snowstorm at the landscape level. **Figure 21** shows the effect of the snow storm within a more localized portion of the forest from a drone image acquired right above the forest canopy height.

Figure 22 is one of the unique abilities of a drone, which is demonstrated by this particular drone image, which is a nadir shot of an individually snow-covered tree to visualize the amount of snow collected on individual tree branches. Video 6 was flown to depict videography of the snow storm using multiple flight heights starting at ground level to visualize the snow storm impact of an East Texas forest.

Figure 21.
Drone image showing localized snow accumulation within a forest.

Figure 22.
Nadir drone image of an individual snow-covered tree.

9. Impacts of a 25-year flood event

Historic flooding can have a severe impact on a local natural ecosystem. The increased water levels and increased water flow rates within streams and rivers can cause havoc to a multitude of natural resource ecosystems. In 2023, Nacogdoches, Texas was inundated with flood water from an historic 25-year flood event. During, and immediately after the flooding event, faculty, and students from SFASU acquired drone imagery to document the floods impact on the SFASU campus and the town of Nacogdoches.

Figure 23, a high oblique drone image of La Nana Creek that flows thru the middle of the SFASU campus was acquired to document the amount of overflow over the

Figure 23.
Drone image showing flooding extent of Lanana Creek.

Figure 24.
Drone image showing impact of historic flood on Starr avenue.

Figure 25.
Drone image showing the extent of flooding on SFASU intramural field.

banks of Lanana Creek. **Figure 24** is a drone image of where Lanana Creek crosses Starr Avenue south of the SFASU campus to aid the City of Nacogdoches to historically document the destructive nature of the flood event.

Figure 25, a high oblique drone image, was acquired to document the amount of flooding within the intramural field on the SFASU campus. Video 7 was flown over Lanana Creek on the SFASU campus to document the flooding of the creek and create videography showing flood damage in and around the SFASU campus in close proximity to Lanana Creek. Video 8 depicts the utility of using drone videography to historically document the damage of an historic flooding event to city road networks.

10. Conclusions

The applicability of using drones to aid individuals involved with natural disasters, such as tornados, high wind damage, lightning strikes, ice storms, snow storms, and historic flood events, has been shown to be an effective tool for a practicing natural resource manager. When a natural disaster occurs, a drone has been shown to be a useful tool to quantify the effect of a natural disaster from a visual perspective. Drones, with their ability to observe the surface of the earth from a distance using either a nadir or oblique perspective can provide individuals involved with disaster management evaluation and clean up effort's information not attainable or not possible with a traditional field-based site visit. The ability to quickly respond to a disaster and capture real-time footage as images, video and processing into both two- and three-dimensional products aids in the resource recovery. In addition, the ability to utilize drone cameras and sensors with zoom capabilities further extends the information gathering efforts of disaster management relief workers. The utility of using drones to monitor and access disasters not specifically related to natural resources and across a more diverse disaster context is implied. The authors, as natural resource managers, wished to focus specifically on disasters related to natural resource ecosystems with specific drone imagery that they have collected to visually enhance the utility of drones within a disaster setting.

Acknowledgements

The authors wish to acknowledge the many students over the years who have helped collect drone imagery and participated in learning how to use drones to map, manage, and monitor forests and natural resources. It is through their valuable contribution and love of learning that has made the validation of introducing drones within ATCFA and demonstrating their use within disaster management scenarios a success. This work was supported by McIntire Stennis funds administered by the Arthur Temple College of Forestry and Agriculture and a Center for Applied Research and Rural Innovation grant from Stephen F. Austin State University.

Conflict of interest

The authors declare no conflict of interest.

Author details

Daniel R. Unger, David L. Kulhavy*, I-Kuai Hung, Victoria Williams, Yanli Zhang and Reid Viegut
Arthur Temple College of Forestry and Agriculture, Stephen F. Austin State University, Nacogdoches, Texas, USA

*Address all correspondence to: dkulhavy@sfasu.edu

References

[1] Bullard SH. Forestry curricula for the 21st century—Maintaining rigor communicating relevance, building relationships. Journal of Forestry. 2015;**113**:552-556. DOI: 10.5849/jof.15-021

[2] Kulhavy DL, Unger DR, Hung I-K, Williams VM, Zhang Y, Viegut R. Integrating drones across the curriculum at Stephen F. Austin State University. IntechOpen; 2023. DOI: 10.5772/intechopen.1002705

[3] Daud SMSM, Yusof MYPM, Heo CC, Khoo LS, Singh MKC, Mahmood MS, et al. Applications of drone in disaster management: A scoping review. Science and Justice. 2022;**62**:30-42. DOI: 10.1016/j.scijus.2021.11.002

[4] Bullard SH, Stephens Williams P, Coble T, Coble DW, Darville R, Rogers L. Producing "society-ready" foresters: A research-based process to revise the Bachelor of Science in forestry curriculum at Stephen F. Austin State University. Journal of Forestry. 2014;**112**:354-360. DOI: 10.5849/jof.13-098

[5] Bullard S, Coble D, Coble T, Darville R, Rogers L, Stephens WP. Producing "Society-Ready" Foresters: A Research-Based Process to Revise the Bachelor of Science in Forestry Curriculum at Stephen F. Austin State University [ATCOFA Monograph 1-2014]. Nacogdoches: Arthur Temple College of Forestry and Agriculture, Stephen F. Austin State University; 2014

[6] Unger DR, Kulhavy DL, Hung I-K, Zhang Y, Williams S. Integrating drones into a natural-resource curriculum at Stephen F. Austin State University. Journal of Forestry. 2019;**117**:398-405. DOI: 10.1093/jofore/fvz031

[7] Zhang Y, Kulhavy D, Gerland J, Hung I, Unger D, Wen X, et al. Evaluating different UAS flight models for 3D model generation and printing of a tornado destroyed cultural heritage: Caddo House in Texas. Drones and Autonomous Vehicles. 2023;**1**:10003. DOI: 10.35534/dav.2023.10003

[8] Leysath M, Galen R. Shahó and the power of place. Teaching Artist Journal. 2021;**19**:1-13. DOI: 10.1080/15411796.2021.1925555

Chapter 5

A Set of Proposals to Overcome the Crisis in Atmospheric Remote Sensing Technologies

Yury Alex Polkanov

Abstract

This chapter summarizes proposals to expand the capabilities of lidar measurements of the Earth's atmosphere as a structural stratified medium with complex thermodynamics. The generalized structure of the revealed, statistically stable horizontal optical inhomogeneities of the surface layer and their dynamics are taken as an informative indicator of the thermodynamic stability of the atmosphere in the zone even beyond the limits of measurements. A special statistical method for analyzing the two identified structures is used. It is based on the separate fixation and statistical averaging of the intervals between local maxima and minima of the lidar backscattering signal and the analysis of the time dynamics of such an interval component of the signal during the night, under anticyclone conditions. It assumed that this approach makes it possible to identify certain statistically stable structures of atmospheric inhomogeneities of varying connectivity, depending on the overall thermodynamic stability of the atmosphere. In addition, a specific method is tested to search for emerging and temporarily stable anomalies of such a structure of atmospheric inhomogeneities, but associated with a violation of the thermodynamic stability of the atmosphere. Such a filtration method (identification of a certain anomaly against the background of an extended organized structure of inhomogeneities) showed high efficiency in model experiments compared to standard filtration methods.

Keywords: lidar measurements, atmosphere, optical inhomogeneities, structure, signs of organization, thermodynamic stability, anomaly, backscattering signal, sampling, temporal organization of radiation and registration

1. Introduction

Classical lidar measurements of the atmosphere assumed obtaining information about the properties of the medium by recording and special processing of the backscattering signal, usually in the photon counting mode. The received signal assumes its digitization and sampling with a given constant resolution. As a result, we get a certain sequence of numeric values that are used in subsequent processing to restore the properties of the medium based on such a signal [1, 2]. This approach to the organization of measurements grew out of laboratory experiments conducted in an environment with stable properties and stable thermodynamics. This standard technology for measuring and processing the received signal leads to an exponential increase in technical problems when trying to develop such an approach. These difficulties led to a reorientation of lidar systems to create models and maps of objects based on a signal directly reflected from the underlying surface. The purpose of the research is to find ways to circumvent the identified problems for the further development of lidar measurements of atmospheric properties directly, to increase their effectiveness by revising the temporary organization of sampling and digitization of the received signal and reorientation to more effective and informative research directions. The complex of ideas was formed by 1989.

One of such directions is the study of the processes of formation of the structure of optical inhomogeneities in the atmosphere. Under certain conditions, the atmosphere is a stratified medium in which the so-called internal gravitational waves arise and propagate [3–5]. The sources are: flows over an inhomogeneous surface, local jumps in the thermodynamic characteristics of the medium (on medium seals, on atmospheric fronts, on sources of turbulence in the boundary layer above the surface, during convection under the temperature inversion layer. Favorable conditions for the formation of such structures arise in the surface layer, especially at night, in anticyclone conditions.

In the presence of surface temperature inversion and extended fields of polluting urban aerosol, at the boundaries of stratification layers, the amplitude of internal waves increases, becoming longer than the wavelength, which leads to loss of stability and collapse of waves. The potential energy of the wave is used to mix the medium in the collapse zone, which leads to the formation of turbulent spots of relative uniformity in density. Moreover, they have sharp boundaries and a long scattering time (scattering velocity ~ 10 cm/s). Stable stratification of the medium helps to suppress turbulence in the flow, but in such spots it exists for a long time. This helps to retain a relatively coarse aerosol in such spots, which usually settles under the influence of gravity, which leads to intense light scattering on such structures. The degree of scattering increases with increasing power of the turbulent structure and with the concentration of the polluting aerosol, which serves as a kind of "developer" of the structure of inhomogeneities.

Indeed, visual [6], radar [7, 8], radiometric [9], lidar [10–13], and radar [14, 15] measurements indicate the presence of similar ordered structures in the atmosphere. In some cases, their length in the horizontal plane reaches tens or even hundreds of kilometers. The time of their existence varies from seconds (for small-sized structures), minutes, to several hours. The size of the registered formations for the surface layer varies from meters to hundreds of meters, and their contrast varies accordingly from percent to tens of percent. A certain condition for the stability of these ordered structures is the stability of the vertical stratification of the medium, which in this case can be maintained in time intervals from several minutes to several days. However, the existing data only allow us to state the existence of a connection, without specifying the conditions of its existence, due to the low efficiency and insufficient information content of the measurements carried out. Local measurements at a limited number of points do not allow us to specify the nature and features of the relationship of the structure of inhomogeneities of atmospheric formations with the meteorological situation and its stability. The assumption that variations in the power of the scattering signal are determined by fluctuations in the attenuation coefficient of backscattered radiation associated with fluctuations in aerosol density in turbulent structures is confirmed by a whole set of data

[5, 6, 16, 17]. This suggests the prospects of using lidar systems for this type of tasks, with a certain agreement between the properties of the equipment and the medium under study.

2. Formulation of the problems

By now, the energy growth of lidar systems is achieved with a significant increase in weight, dimensions and a clear violation of environmental safety. With the radiation energy unchanged, a decrease in the measurement error is accompanied by a deterioration in the spatial resolution along the sounding path. Increasing the accuracy of measurements has limitations due to the temporal variability of the medium itself. Under these conditions, attempts to reduce the measurement error of the signal by increasing the signal registration gate or by accumulating the signal in a series of measurements are blocked by a concomitant deterioration in spatial and temporal resolution. Such limitations are caused by a number of implicit contradictions present in the technologies of remote measurement and processing of the obtained results, which have escaped the attention of researchers.

2.1 Conflict 1

The confrontation between the zone of ongoing large-scale processes under study and the desire for hardware sampling of the incoming signal with increasing resolution. This excludes the assumption of the medium as a coherent whole formation, with obviously complex thermodynamics, and reduces the entire analysis of the properties of a specific, virtually isolated point of the medium (in the extreme case).

2.2 Conflict 2

The confrontation between the requirement to increase the required energy of the pulsed laser used, to increase both the measurement range and the spatiotemporal resolution of the recorded signal while reducing the environmental safety of the measurements. The harmful effects on natural processes are increasing and the danger of irreversible effects on humans and wildlife is increasing.

2.3 Conflict 3

The confrontation between the orientation to measuring the properties of the medium at a point and volumetric thermodynamic processes that affect the overall organization and stability of the spatial and temporal structure of the inhomogeneities of the medium.

2.4 Conflict 4

The confrontation between the need to smooth out signal changes caused by the structure of its inhomogeneities, background noise, and the need to amplify signal changes caused by the appearance of abnormal inhomogeneity in the medium (not peculiar to its equilibrium state).

3. Overcoming technologies

3.1 Technology 1

Overcoming the *first contradiction* involves changing the sampling algorithm of the incoming backscattering signal and its subsequent digitization. Here, the concept of sampling with variable duration is introduced, when the minimum duration of the reference corresponds to the time when the radiation pulse passes through half of the probing path in the zone of single scattering, when two sets of samples are formed [18]. The first set is formed with a constant initial boundary and the growth of each subsequent sample by the required sampling step for the far half of the route. The second set is formed accordingly with a constant final boundary for the near half of the route. Similarly, the second sequence of samples from the far half of the sensing path is organized accordingly by forming an initial sample from the scattering signal of the probing signal from the far half of the sensing path (**Figure 1**).

Moreover, the length of the route corresponds to the zone of single signal scattering and provides a signal-to-noise ratio of at least the selected one in each sample, and the noise consists of external illumination and internal noise of the equipment itself. Similarly, the second sequence of counts of the far half of the route is formed by forming an initial count from the scattering signal of the probing signal from the far half of the route (**Figure 1**).

The rationale for the chosen discretization is based on numerical simulations based on real measurements with a constant sampling step carried out in the laboratory of condensed matter at the Belarusian State University [3–5]. The results of such a comparative experiment [6] are shown in the **Figure 2**.

The presented graph is illustrative. Eight measurement error values are given, normalized for the error at the end of the selected sounding route for the old system, with a characteristic atmospheric attenuation coefficient of 0.3 km^{-1} and a uniform step along the route. The specific quantitative characteristics of such modeling are presented in detail in Refs. [17, 18].

For the proposed sampling technology of the incoming signal, the error varies slightly from strobe to strobe over most of the route, which makes it possible to reduce hardware requirements due to the small dynamic range of the recorded signal.

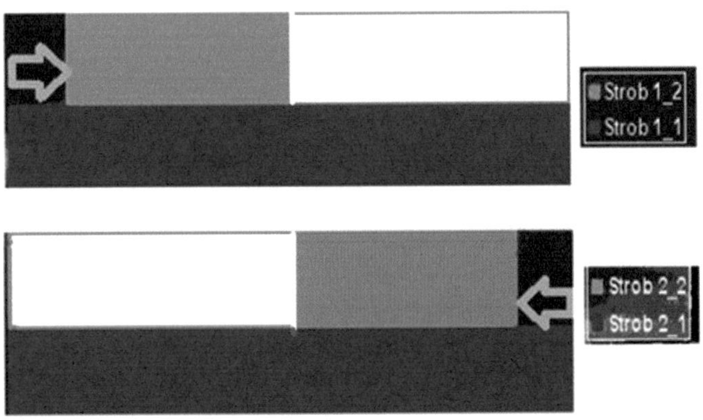

Figure 1.
The principle of the temporary organization of double (counter) discretization of samples of the backscattered lidar signal.

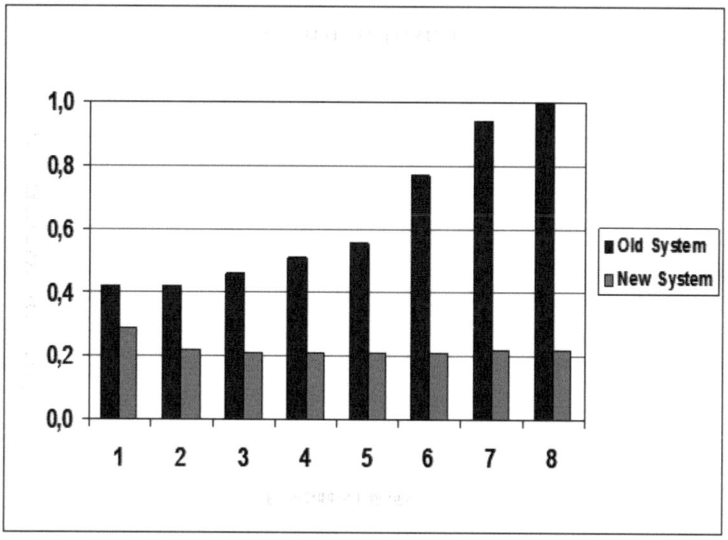

Figure 2.
The dependence of the normalized measurement error on the length of the measurement zone for registration gates or pulses of probing radiation commensurate with the length of the measurement zone (new system) *in relation to the conventional, standard system* (old system).

The restoration of the characteristics of the medium with the required resolution is achieved by sequentially comparing neighboring pairs of samples having a constant initial (final) boundary of each sample and differing in the final boundaries for the selected sampling step (see **Figure 1**).

The simulation results showed that the proposed measurement scheme makes it possible to carry out measurements at a high level of daytime illumination conditions, while maintaining an error in the range of ~1 ... 10% [19]. In addition, the approach used makes it possible to shift the task of improving measurement accuracy from an area related to the environment to an area related only to hardware capabilities, to an area more subject to control.

3.2 Technology 2

Overcoming the *second contradiction* involves the transition from using a high-power pulsed laser to using a continuous radiation laser of significantly lower power. For this purpose, it is proposed to use the zone of a single interruption of radiation for a time close to the time of radiation of a pulsed laser of a conventional lidar system as a probing "dark" pulse [20, 21]. Parallel modeling of measurements for such a system has shown that the above conclusions on the resolution of the first contradiction are applicable in this case. Continuous 0.5 watt laser radiation can provide the necessary sensing ranges, measurement accuracy and resolution along the sensing path. Combining both technologies for both the radiating part of the equipment and its recording part will allow for even better results.

3.3 Technology 3

Overcoming the *third contradiction* involves shifting the focus of attention from classical studies of properties at each point to the medium as an extended object with a certain stratification, the properties of which depend on the stability of its thermodynamic state. As such a criterion, the generalized value of the difference between the given Richardson numbers taken, which was calculated according to a certain method based on meteorological parameters obtained from the nearest weather station [18]. The search for signs of organization in the structure of the recorded signal and its correlation with the structure of the real environment, its variability and restructuring associated with a violation of the stability of the thermodynamic state not only in the measurement zone but also beyond it, comes to the fore.

Lidar studies of the atmosphere at night in the surface layer under conditions of a fairly stable anticyclone [19] have shown that the structure of horizontal optical inhomogeneities is quite regular [22, 23]. In some cases, this is clearly observed as the collapse of so-called internal waves (such as Kelvin-Helmholtz waves). A clear example of this situation is presented below (**Figure 3**).

Lidar sensing makes it possible to register a backscattering signal from the atmosphere as the radiation pulse passes through the selected sensing path. Subsequent processing of such a signal using a special technique allows us to uniquely correlate the structure of the transformed signal with the structure of atmospheric inhomogeneities [24]. The complex nature of such structures and their dynamics forced us to abandon the initial harmonic analysis, exclude the amplitude characteristic of the signal from consideration and switch to an independent analysis of the intervals between local maxima or minima of the transformed signal using statistical analysis. Previously, a model experiment was conducted in which the harmonic signal was distorted in a certain way. In harmonic analysis, such distortions appear as a weak expansion of the peak of the corresponding correlation function (when compared with an accurate sine wave). In our case, such distortion manifests itself as a clear additional peak, which indicates a high probability of detecting such distortions in a fairly simple way (**Figure 4**).

In the case of processing real lidar signals, the ultimate goal was to obtain a model in the form of a generalized interval structure of inhomogeneities [11, 12]. The mathematical expectations of the *intervals* between local maxima (*M1*), local minima (*M2*), maxima and minima (*M3*), as well as their variance were taken as characteristics of the identified structures (**Figure 5**).

Figure 3.
Clouds along the horizon in Birmingham, Alabama, December 16, 2011 (image credit: ABC 33/40 in Birmingham).

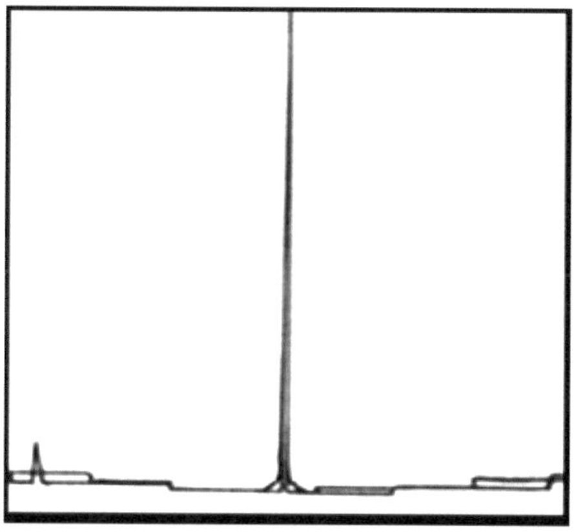

Figure 4.
Normalized cross-correlation function of power spectra or time interval distributions between signal extremes for accurate and distorted sinusoids, respectively (numerical simulation).

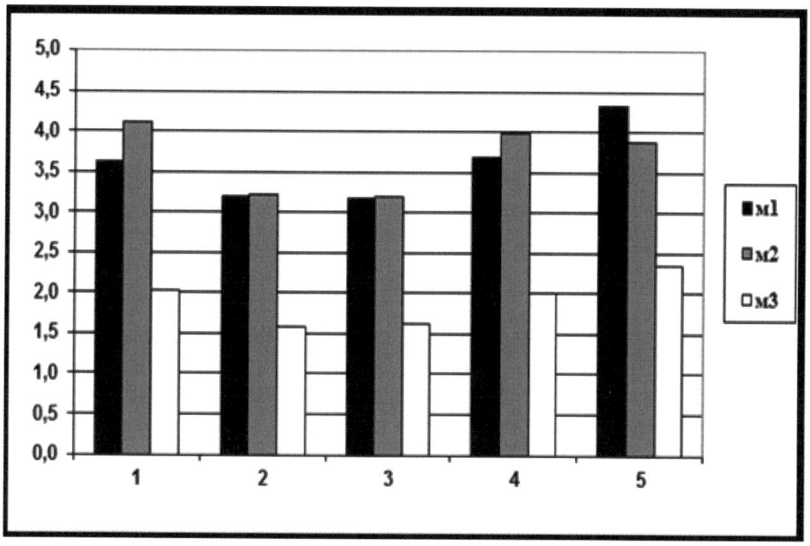

Figure 5.
*The results of determining the generalized interval between the elements of the structure of different types (*M1, M2, M3*), for the optical heterogeneities of the atmosphere in the* five *selected sensing areas.*

As can be seen, there is a changing difference between the "positive" (*M1*) and "negative" (*M2*) structure of inhomogeneities and the intervals between them (*M3*). In addition, such an analysis allows us to relate the features of the detected structures to the thermodynamic stability of the atmosphere. To do this, we used data on meteorological parameters from the nearest weather station. The generalized Richardson number [13] was used as a criterion for the thermodynamic stability of the medium, which was calculated on the basis of the mentioned meteorological parameters. It turned out that the structure of inhomogeneities becomes more complicated with a decrease in the stability of the atmosphere. There is a transition from a simple regular structure of inhomogeneities, the appearance of long-wave component, through an intermediate stage of growth of the dispersion of the initial structure [24–27]. Moreover, the statistical relationship between the "positive" and "negative" structure decreased with decreasing thermodynamic stability of the atmosphere. They became more and more independent, which is obviously not revealed by standard Fourier analysis methods.

3.4 Technology 4

Overcoming the *fourth contradiction* involves the use of a nonlinear signal filtering method, when, after sampling and digitization, its periodic inhomogeneities are significantly smoothed out, and one-time anomalies are amplified. Numerical modeling was carried out using an interactive system of spectral statistical processing of a noisy model signal. An exponentially decreasing signal with a drop in the center of the sample of 50 samples was used [28, 29]. The effectiveness of such filtration has been verified in a numerical experiment, when compared with linear and median filtration. An example of such filtering of a model signal is shown in **Figure 6**.

Filtration allows you to clearly identify the boundary of the difference. The signal after the drop exceeds the original value. In this case, linear and median filtering are not effective and do not allow identifying the introduced drop in the model signal (**Figure 7**).

It should be expected that the transition from an anticyclone-type situation to a cyclone will cause a significant transformation of the structure of heterogeneities, even before the appearance of obvious thermodynamic signs of such a transition. The conducted studies suggest that the occurrence of thermodynamic instability of the atmosphere, even outside the observation zone, leads to a proactive sharp violation of the stability of the structure of optical heterogeneities of the atmosphere, long before a visible significant change in the situation in the observation zone. Predictors of such a change may be some kind of soliton-type anomalies and can be clearly identified by the described filtering.

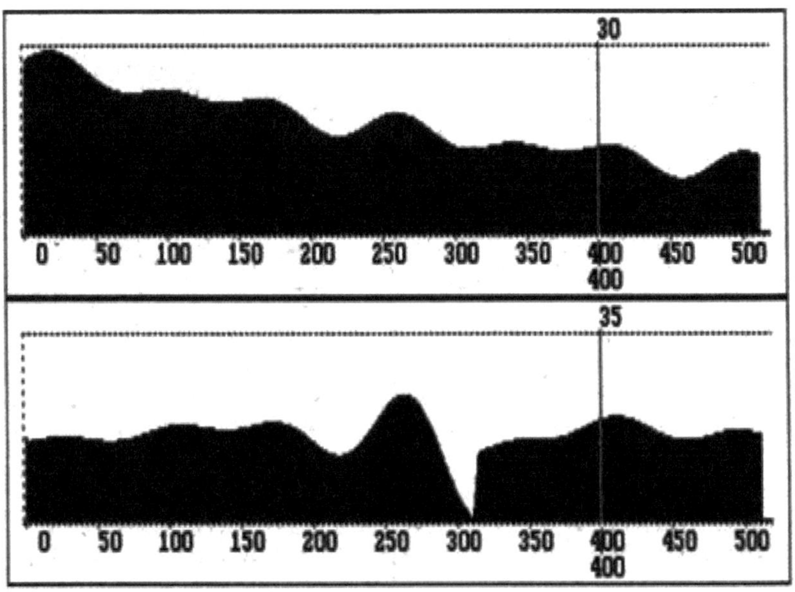

Figure 6.
A model signal with an introduced difference at the noise level and an imposed structure (upper graph) and filtering results revealing such a drop (lower graph), where the ordinate axis is the magnitude of the signal, and the abscissa axis is the number of samples.

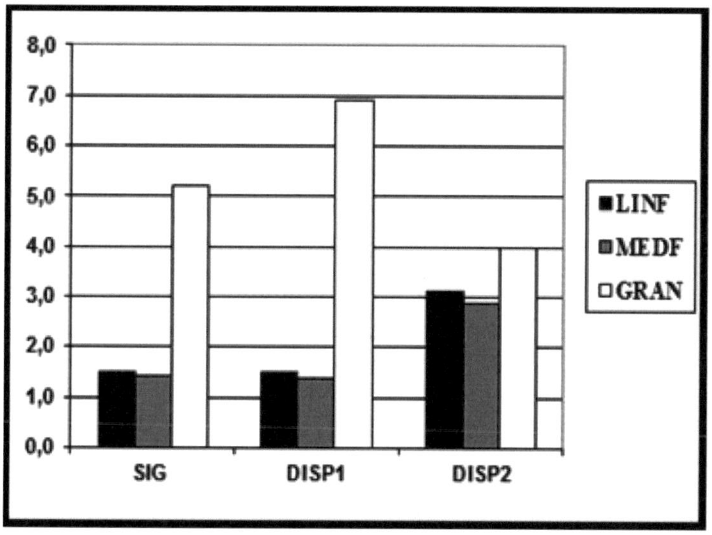

Figure 7.
Results of comparison of the proposed processing of the incoming scattering signal (nonlinear filtering—GRAN) with linear and median filtering methods (LINF, MEDF). DISP1, DISP2—dispersion before and after the introduced signal difference, SIG1, 2, 3—signals at the ratio between the maximum value, drop and noise: 10:1:1; 10:1:5; 10:1:10, respectively.

3.5 Technology 5

One of the visual options for representing a meteorological situation may be the creation of a generalized model structure of inhomogeneities (based on the results of a large-scale array of real measurements) as a basis for creating a visual image of a developing nonlinear process in a continuous environment, as an integral object.

A structure such as the surface of a model material (similar to the structure of a hologram) can be used as a holographic visual image of the process. The final stage of the proposed technology may be the use of information about the unit cell of the structure of inhomogeneities to create a kind of model of the surface of the material with a structure similar to that of a hologram. The dynamics of the generated holographic images of the structure of inhomogeneities will allow us to visualize the process of changing the stability of the situation. The more catastrophic the change, the more complex the recorded structure and the formed holographic image acquires greater clarity. You can even use the symbolic name of such an image - "Smile of the Devil". His "facial expressions" will allow, without going into complex mathematical operations, to quickly fix the degree of catastrophism of developing processes. One of the visual options for representing the meteorological situation may be the creation of a generalized model structure of heterogeneities (based on the results of a volumetric array of real measurements) as a basis for creating a visual image of a developing nonlinear process in a continuous environment, as an integral entity.

4. Conclusion

The proposed set of technologies for measuring, recording, and processing remote measurements makes it possible to create lidar sensors for monitoring the stability of a meteorological situation and use them in a network of standard weather stations.

Such technologies are applicable for a wide range of tasks for processing noise-like signals from various continuous media, where it is possible to relate the complexity of the structure of heterogeneities to the "thermodynamic" stability of the situation, for example, in the banking sector or sociological research.

Author details

Yury Alex Polkanov
Independent Researcher, Minsk, Belarus

*Address all correspondence to: ypolk@yandex.ru

References

[1] Vladimirovich ZV. Lidar Control of the Stratosphere. Novosibirsk: Ross. Acad. Sci. Sib. Department Institute of Atmospheric Optics; 2004. 306 p

[2] Bezrukova NA, Chernokulsky AV, Izvestiya RAS. Russian studies of clouds and precipitation in 2015-2018. Physics of Atmosphere and Ocean. 2020;**56**(4):397-417. DOI: 10.31857/S0002351520040021

[3] Gossard E, Hook W. Waves in the Atmosphere, Mir, 1978. 532 p

[4] Lighthill D. Waves in Liquids. Mir, 1981. 598 p

[5] Barenblatt GI, Monin AS. On the possible mechanism of the phenomenon of discoid formations in the atmosphere. Doklady USSR Academy of Sciences. 1979;**246**(4):834-837

[6] Grechko GM, Gurvich AS, Romenenko YV. The structure of density inhomogeneities in the stratosphere according to observations from the Salyut 6 orbital station. Izvestia of the USSR Academy of Sciences, FAO. 1980;**16**(4):339-344

[7] Gjssard EE, Richer JH. View of internal waves of limited amplitude as measured by high-resolution radar. Journal of the Atmospheric Sciences. 1970;**27**:971-973

[8] Philip MT, Kent GS. Short-period temporal variations of the atmospheric aerosol layer. In: 9-th Int. Laser Nadar Conf. Laser Atmos. Stud; Munich. 1979. pp. 161-162

[9] Jonson WB, Evans WE. Lidar measurements of the diffusion of plumes from high pipes. In: 3-rd Conference on Laser Radar Studies of Atmosphere Conference Abstracts; Jamaica. 1970. p. 12

[10] Voevodin YM, Zadde GO, Matvienko GG, Samokhvalov IV. Spatial inhomogeneities of the backscattering coefficient according to lidar sensing data. In: Issues of Laser Sensing of the Atmosphere. Nauka; 1976. pp. 45-53

[11] Werner C. Lidar measurement of atmospheric aerosol as a function of relative humidity. Opto-Electronics. 1972;**4**(2):125-132

[12] Voevodin YM, Zadde GO, Matvienko GG, Samokhvalov IV. Investigation of aerosol statistical characteristics based on laser sensing data. In: XI All-Union Conference on the propagation of Radio Waves. Tez. dokl., part II; Kazan. 1975. pp. 116-118

[13] Golev IN, Parfenov OP, Kaprielov VN, Ivlev IK. Lidar observations of internal gravitational waves. Doklady Bolgarskoj Akademii Nauk. 1986;**39**(1):43-46

[14] Kalistratova MA, Nesterova TN, Petenko IV, Smirnov AS, Time NS, Shurygina EA. Measurement of statistical characteristics of an echo signal during acoustic sounding of the atmosphere. Izvestia of the USSR Academy of Sciences, FAO. 1986;**22**(9):987-990

[15] Kalistratova MA, Petenko IV, Shurygin EA. Sodar studies of the wind velocity field in the lower troposphere. Izvestia of the USSR Academy of Sciences, FAO. 1987;**23**(5):451-461

[16] Goldshtik MA, Stern VN. Structural turbulence in dissipative systems. Preprint 47-81. Novosibirsk; 1981. 47 p

[17] Polkanov YA. Identification of Self-Organization Processes of Continuous

Medium Inhomogeneities. Using the Example of Lidar Measurements of the Atmosphere. LAP Lambert Academic Publishing; 2012

[18] Polkanov YA. Looking at remote sensing the timing of an organisation's point of view and the anticipation of today's problems. In: Remote Sensing of Planet Earth. Yann Chemin Publisher; 2012. Available from: https://www.intechopen.com/chapters/27100

[19] Polkanov YA. The possibility of improving the accuracy of the measurement of a light-scattering signal in the atmosphere. Radiotekhnika i Elektronika. 1988;**33**:2599-2603. Available from: http://adsabs.harvard.edu/abs/1988RaEl...33.2599P

[20] Polkanov YA. Sounding of the environment by means of the un-impulse of the low-power continuous source. In: Singh UN, Pappalardo G, editors. Proceedings of SPIE – Volume 6750 Lidar Technologies, Techniques, and Measurements for Atmospheric Remote Sensing III. 2007

[21] Polkanov YA. Remote sensing of the self-organizing systems by dark. Pulse International Journal of Remote Sensing Applications. 2012;**2**(3). Available from: https://issuu.com/sep2011--now/docs/ijrsa10050-20120926-105928-6154-242

[22] Polkanov YA. Regularity of the structure of the heterogeneities in the scattering of optical radiation and their connection with meteorological parameters. Proceedings of the USSR Academy of Sciences. Physics of the Atmosphere and Ocean. 1985;**21**(7):720-726

[23] Polkanov YA. On the timing of the signal to the structure of the inhomogeneity of an extended medium. Electronic Periodic Peer-Reviewed Scientific Journal. 2022;**110**:63. Available from: https://sci-article.ru/stat.php?i=1666770216

[24] Polkanov YA, Kudinov VN. One possibility of analyzing the periodic structure of a complex signal. News from universities. Instrumentation. 1989;**32**(4):6-11

[25] Polkanov YA. Dynamics of the structure of the scattering signal and its connection with the weather situation. Proceedings of the USSR Academy of Sciences. Physics of the Atmosphere and Ocean. 1989;**25**(6):599-603

[26] Polkanov YA. Structural-statistical frequency analysis of time series. Electronic Periodic Peer-Reviewed Scientific Journal. 2022;**110**:22. Available from: https://sci-article.ru/stat.php?i=1665135403

[27] Polkanov YA. Consistency of changes in the structure of optical inhomogeneities of the atmosphere and a set of meteorological parameters. Meteorology and Hydrology. 1991;**3**:39-48. Available from: http://cat.inist.fr/?aModele=afficheN&cpsidt=5237241

[28] Polkanov YA. A method for determining the position of optical inhomogeneity of the atmosphere. Auto. USSR certificate N1448907, class. G01W1/00 dated 06/01/86, publ. 09/01/88

[29] Polkanov YA. One possibility of identifying anomalous heterogeneity of the atmosphere (nonlinear filtering method). Optics of the Atmosphere and Ocean. 1992;**5**(7):720-772

Section 2

Applications and Case Studies

Chapter 6

Earth Observation Applications to Agricultural and Environmental Statistics

Luis Ambrosio, Luis Iglesias, Carmen Marin, Sophie Bontemps, Böris Nörgaard, Pierre Houdmont, Cosmin Cara, Cosmin Udroin, Laurentiu Nicola, Sadri Haouet and Zoltan Szantoi

Abstract

Pixels classifiers do not allow assessment of uncertainty due to classification errors. This is a major limitation to the direct use of crop type maps (CTM) as agricultural or environmental statistics. Here, we include the results of our recent research to overcome this limitation, using statistical models that integrate ground and CTM data to evaluate any source of uncertainty, including classification errors. We use linear models for field data that can be treated as values of a continuous variable, such as crop acreage observed at parcel level, and multinomial logit models for field data that can be treated as values of a categorical variable, such as the type of crop observed at the pixel level. Among the tasks required to produce agricultural and environmental statistics, we focus on three for which CTM are especially useful: improving the effectiveness of currently used methods, increasing the granularity of statistics by producing small area estimations, and optimizing the sample design. A prototype for integrating ground and CTM data to achieve each of them is included, together with the results of its application to data provided by two National Statistical Offices.

Keywords: crop type maps and ground data integration, linear models, multinomial logit models, efficiency, spatial resolution

1. Introduction

"End hunger, achieve food security, and promote sustainable agriculture" is the second (the first is "end poverty") Sustainable Goal in the United National 2030 Development Agenda. To implement policies and practices in pursuit of these goals, and to monitor and control such policies, detailed information about land uses, crops acreage, and crops yield is necessary.

International programs to improve agricultural and environmental statistics have been recently launched by the World Bank and the Food and Agriculture Organization

(FAO) of the United Nations. Looking for synergies with these programs, and with the aim of facilitating the use of remote sensing (RS) information at the National Statistical Offices (NSOs) supporting agricultural and environmental statistics, the European Space Agency (ESA) launched the Sentinels for Agricultural Statistics (Sen4Stat) project.

In this chapter, along with well-established results in the RS literature (the statistical classifiers for crop type maps (CTM) generation), we include the main results of our recent research in the applications of CTM to agricultural and environmental statistics, carried out within the framework of Sen4Stat and partially published in the Statistical Journal of the IAOS [1]. To date, pixel classifiers do not allow the assessment of uncertainty due to classification errors. This is a major limitation to directly using CTM data as statistical estimates, and here, we focus on how to use statistical models to integrate CTM with ground data to produce efficient agricultural and environmental statistics, taking into account any source of uncertainty, including the classification errors.

In large areas, where the sample size for collecting ground data is large enough, we follow an approach based on designing a probabilistic scheme to select the sample and on statistical models to integrate the ground data with CTM data. If ground data can be treated as values of a continuous variable, such as crop acreage observed at the parcel level, we use linear models. If ground data can be treated as values of a categorical variable, such as the type of crop observed at the pixel level, we use multinomial logit models. This design-based approach is robust in the sense that the inference is based not on the model, but on the sampling design, and the estimates are design-consistent. In small areas, where the sample size is small, the design-based approach is not accurate enough for most applications, and the inference is based on the model.

We focus on three of the tasks necessary to produce agricultural and environmental statistics, for which CTM is especially useful: improving the efficiency of currently used methods, increasing the granularity of statistics, and optimizing sample design. A ground and CTM data integration prototype to conduct each of these tasks is included, along with examples of its application to data provided by National Statistical Offices.

2. Pixels classification error

Many RS applications involve the classification of the $i = 1, 2, \cdots, N$ population pixels, using multispectral (L bands) reflectance data, $\mathbf{x}_i = \underset{1 \leq l \leq L}{row}(x_{li})$ [2]. In agricultural and environmental statistics, the product of pixels' classification most often used is a CTM, where a class is associated with each one of the type of crops or land uses, $k = 1, 2, \cdots, K$.

However, pixels classifiers are subject to error, and as a result, CTM are subject to uncertainty. In this section, we will show that pixel classifiers do not allow the assessment of this uncertainty. This is a major limitation to use directly CTM as agricultural or environmental statistics, because in addition to estimates of the study population characteristics, measures of the uncertainty of these estimates are required.

2.1 Classifiers

Let $i \in k$ denote the event "pixel i belongs to class k." Except for the few pixels included in the sample, there is uncertainty about the occurrence or not of this event, and to model this uncertainty, as well as the instrument measurement errors, we consider the joint probability function, $f(i \in k, \mathbf{x}_i)$, of the two events "pixel i belongs to class k and its spectral measure is $\mathbf{x}_i$." This function is used to define the decision rule: "Classify i in the class k for which the probability $f(i \in k, \mathbf{x}_i)$ is maximum: $f(i \in k, \mathbf{x}_i) = \max_{k=1,2,\cdots,K} f(i \in k, \mathbf{x}_i)$." Three main approaches proposed in the literature to implement this rule are (i) maximum likelihood, (ii) maximum entropy, and (iii) multinomial regression.

The maximum likelihood (ML) approach [2] assumes that $\mathbf{x}_i$ is distributed, within a given class k, according to a multivariate normal, $\mathbf{x}_i | i \in k \rightarrow N_L(\mathbf{m}_k, \mathbf{V}_k)$, of mean $\mathbf{m}_k = E(\mathbf{x}_i | i \in k)$, and covariance matrix $\mathbf{V}_k = Var(\mathbf{x}_i | i \in k)$. Under this assumption, the decision rule $f(i \in k, \mathbf{x}_i) = \max_{k=1,2,\cdots,K} f(i \in k, \mathbf{x}_i)$ is equivalent to classify i in the class k if $g_k(\mathbf{x}_i) = \max_{k=1,2,\cdots,K} g_k(\mathbf{x}_i)$, where $\{g_k(\mathbf{x}_i) = \log f(i \in k) - \frac{1}{2}\log|\det \mathbf{V}_k| - \left[\frac{1}{2}(\mathbf{x}_i - \boldsymbol{\mu}_k)\mathbf{V}_k^{-1}(\mathbf{x}_i - \boldsymbol{\mu}_k)^T\right]; i = 1, 2, \cdots, N; k = 1, 2, \cdots, K\}$ (see Appendix 1). $\mathbf{m}_k$ and $\mathbf{V}_k$ are unknown but can be estimated using a simple sample of n_k pixels and the estimators $\hat{\mathbf{m}}_k = \underset{1 \le l \le L}{col}(\hat{m}_{lk}) = \underset{1 \le l \le L}{col}\left(\frac{1}{n_k}\sum_{i=1}^{n_k} x_{lki}\right)$ and $\hat{\mathbf{V}}_k = \underset{1 \le l \le L}{col}\ \underset{1 \le l' \le L}{row}\left(\frac{1}{n_k - 1}\sum_{i=1}^{n_k}(x_{lki} - \hat{m}_{lk})(x_{l'ki} - \hat{m}_{l'k})\right)$.

Two disadvantages of this approach are that normality is a very restrictive assumption for reflectance data and that it does not allow a direct computation of class probability, $\mu_{ik} = P(i \in k | \mathbf{x}_i)$ (given $\mathbf{x}_i$, μ_{ik} is the probability that crop k covers pixel i).

The maximum entropy (ME) approach has the advantage that it does not require any assumptions about statistical distribution and directly provides estimates of the probability of each class. The entropy is an uncertainty measure: for the ith pixel, the quantity of uncertainty in $\boldsymbol{\mu}_i = \underset{1 \le k \le K}{col}(\mu_{ik})$ is $H(\boldsymbol{\mu}_i) = -\sum_{k=1}^{K}\mu_{ik}\ln\mu_{ik}$. To estimate $\boldsymbol{\mu} = \underset{1 \le i \le N}{col}\ \underset{1 \le k \le K}{col}(\mu_{ik})$, we use crops data observed in a sample of n pixels, $\mathbf{y} = \underset{1 \le i \le n}{col}\ \underset{1 \le k \le K}{col}(y_{ik})$ where $y_{ik} = 1$ if crop k covers pixel i and $y_{ik} = 0$ otherwise, together with RS data $\mathbf{X} = \underset{1 \le i \le n}{col}\ \underset{1 \le l \le L}{row}(x_{li})$. The maximum entropy principle is to choose the class probability values, μ_{ik}, compatible with the observed data, $\mathbf{y}$, that makes $H(\boldsymbol{\mu}) = -\boldsymbol{\mu}^T\ln\boldsymbol{\mu} = -\sum_{i=1}^{n}\sum_{k=1}^{K}\mu_{ik}\ln\mu_{ik}$ maximum.

The condition of compatibility between $\boldsymbol{\mu}$ and $\mathbf{y}$ is specified by the model $\mathbf{y} = \boldsymbol{\mu} + \boldsymbol{\varepsilon}$, where $\boldsymbol{\varepsilon} = \underset{1 \le i \le n}{col}\ \underset{1 \le k \le K}{col}(\varepsilon_{ik})$ and ε_{ik} is an unobserved random perturbation. The RS data are introduced by transforming this model into a set of consistency restrictions: $(\mathbf{I}_K \otimes \mathbf{X}^T)\mathbf{y} = (\mathbf{I}_K \otimes \mathbf{X}^T)\boldsymbol{\mu} + (\mathbf{I}_K \otimes \mathbf{X}^T)\boldsymbol{\varepsilon}$, to which we impose the condition of uncorrelation between spectral measures and random perturbations, $(\mathbf{I}_K \otimes \mathbf{X}^T)\boldsymbol{\varepsilon} = \mathbf{0}$, to reduce the consistency restrictions to $(\mathbf{I}_K \otimes \mathbf{X}^T)\mathbf{y} = (\mathbf{I}_K \otimes \mathbf{X}^T)\boldsymbol{\mu}$.

The number nK of unknown class probabilities, $\boldsymbol{\mu} = \underset{1 \le i \le n}{col}\ \underset{1 \le k \le K}{col}(\mu_{ik})$, is higher than the number LK of consistency restrictions, and as a result, there is not a unique solution for $\boldsymbol{\mu}$. Among the countless solutions, we choose the one making $H(\boldsymbol{\mu})$ maximum, subject to the consistency restrictions and the adding-up normalization conditions, $\underset{n}{diag}(\mathbf{1}_K^T)\boldsymbol{\mu} = \mathbf{1}_n$, where $\mathbf{1}_K$ is a $(K \times 1)$ vector of ones, and $\mathbf{1}_n$ is a $(n \times 1)$ vector of ones. This solution is found by solving the Lagrange function

$\Psi = -\boldsymbol{\mu}^T \ln \boldsymbol{\mu} + \beta^T \left[\left(\mathbf{I}_K \otimes \mathbf{X}^T \right) \boldsymbol{\mu} - \left(\mathbf{I}_K \otimes \mathbf{X}^T \right) \mathbf{y} \right] + \lambda^T \left(\mathbf{1}_n - \underset{n}{diag} \left(1_K^T \right) \boldsymbol{\mu} \right)$, where $\boldsymbol{\beta} = \underset{1 \le k \le K}{col} (\boldsymbol{\beta}_k) = \underset{1 \le k \le K}{col} \underset{1 \le l \le L}{col} (\beta_{kl})$ and $\lambda = \underset{1 \le i \le n}{col} (\lambda_i)$.

The multinomial (MNL) approach is similar to the maximum entropy. In fact, the entropy $H(\boldsymbol{\mu})$ can be derived from the multinomial distribution (see Appendix 1). Both approaches directly provide coincident class probabilities estimates, but the former requires to specify a function relating these probabilities with the RS data, and we consider the logit function, $\mu_{ik} = \frac{e^{\sum_{l=1}^{L} \beta_{kl} x_{li}}}{1 + \sum_{k=1}^{K-1} e^{\sum_{l=1}^{L} \beta_{kl} x_{li}}}$, where the coefficients $\boldsymbol{\beta} = \underset{1 \le k \le K}{col} (\boldsymbol{\beta}_k) = \underset{1 \le k \le K}{col} \underset{1 \le l \le L}{col} (\beta_{kl})$ coincide with the Lagrange parameters linked to the consistency restrictions in the ME approach. Consequently, the class probability estimates $\hat{\mu}_{ik}$ based on ME coincide with those based on MNL [3]: for any $i = 1, 2, \cdots, N$ it is $\hat{\mu}_{ik} = \frac{e^{\sum_{l=1}^{L} \hat{\beta}_{kl} x_{li}}}{1 + \sum_{k=1}^{K-1} e^{\sum_{l=1}^{L} \hat{\beta}_{kl} x_{li}}}$ for $k = 1, 2, \cdots, K-1$ and $\hat{\mu}_{iK} = \frac{1}{1 + \sum_{k=1}^{K-1} e^{\sum_{l=1}^{L} \hat{\beta}_{kl} x_{li}}}$, where $\hat{\boldsymbol{\beta}} = \underset{1 \le k \le K}{col} \underset{1 \le l \le L}{col} \left(\hat{\beta}_{kl} \right)$ is the $\boldsymbol{\beta}$ estimate.

Although MNL has the disadvantage over ME that it requires specifying a functional form to link μ_{ik} and $\mathbf{x}_i$, it has the advantage over ML and ME that it allows to calculate the accuracy of the estimates by integrating ground and RS data (even in the case where the classifier used is ML or ME), taking into account the uncertainty due to classification errors.

2.2 Classification errors

The decision rule is subject to errors, and as a result, the classification is subject to uncertainty. Let $i \subset k$ denote the decision to "include pixel i in class k." This decision is subject to two types of error: the type one, called false negative and denoted by $i \not\subset k | i \in k$, consists of excluding from a given class k a pixel i that is actually of that class, and the error of type two, called false positive and denoted by $i \subset k | i \notin k$, consists of including a pixel i in a given class that is not of that class.

The computation of the classification error probabilities $P(i \subset k | i \notin k)$ and $P(i \not\subset k | i \in k)$ is almost never possible, mainly due to the high dimension L of RS data [2]. In effect, $P(i \subset k | i \notin k) = P(i \subset k)[1 - P(i \in k | i \subset k)]/[1 - P(i \in k)]$ and $i \subset k \Leftrightarrow \hat{\boldsymbol{\beta}}_k - \hat{\boldsymbol{\beta}}_{k'} > 0; k \ne k'$ so that $P(i \subset k) = P\left(\hat{\boldsymbol{\beta}}_k - \hat{\boldsymbol{\beta}}_{k'} > 0 \right); k \ne k'$. In the MNL approach, the $\hat{\boldsymbol{\beta}}_k - \hat{\boldsymbol{\beta}}_{k'}$ distribution is asymptotically normal L-dimensional, and the computation of $P\left(\hat{\boldsymbol{\beta}}_k - \hat{\boldsymbol{\beta}}_{k'} > 0 \right); k \ne k'$ is complex due to the high L values.

Instead of $P(i \subset k | i \notin k)$ and $P(i \not\subset k | i \in k)$, in practice, several ratios are calculated using the sample data (or, more often, by dividing the sample data into two groups, one for training and the other for testing the classifier), but this does not allow us to evaluate the uncertainty due to classification errors. For instance, if we denote by $n_{i \in k}$ the number of sample pixels that actually are of class k, and by $n_{i \in k \cap i \subset k}$ those among $n_{i \in k}$ that are correctly classify in k, then $n_{i \in k \cap i \subset k} / n_{i \in k}$ is an estimate of the so-called producer's accuracy for class k. In the same way, if we denote by $n_{i \subset k}$ the number of sample pixels included in the class k, then $n_{i \in k \cap i \subset k} / n_{i \subset k}$ is an estimate of the so-called user's accuracy for class k. An estimate of the so-called overall accuracy is $\sum_{k=1}^{K} n_{i \in k \cap i \subset k} / \sum_{k=1}^{K} n_{i \in k}$.

Accounting for the uncertainty due to classification errors is key, and that is why the data resulting from classification cannot be used directly as agricultural or environmental statistics. To take into account any uncertainties, including those due to classification errors, the CTM are integrated with the ground data, using statistical models in the way we show in the following sections.

3. Design-based approach: large area estimation

The approach to producing official statistics is based on a probabilistic scheme designed to select the sample from which to collect the required ground data. This scheme allows us to assign a probability of inclusion in the sample to each population unit. To estimate the characteristics of the study population, we use design-consistent estimators whose distribution is tightly concentrated around the true characteristics value, if the sample size is large enough. The uncertainty of these estimates is assessed using the statistical distribution defined by the inclusion probabilities.

The sample is selected using an area frame or a list frame [4–6]. An example of an area frame is that used by Spain: the national topographic map divided into sampling units, each of which is a square of 700 m on each side (49 hectares). The inclusion probabilities is equal to n/N, where n and N are the number of sampling units in the sample and in the population, respectively [6]. France and the United States also use area frames but using points as sampling units, rather than polygons. France to estimate crop acreage [7], and the United States to collect environmental data and develop a national inventory of natural resources [8, 9].

An example of list frame is that used by Senegal: it is based on the 2013 population census [10]. The sampling unit is a household with agricultural activities. The sampling scheme consists of two stages. In the first stage, a sample of Enumeration Areas (EA) is selected with replacement and probabilities proportional to the number of households in the EA. In the second stage, from each selected EA, a sample of households is selected without replacement and with equal probability among households with agricultural activities. To integrate the data collected on the household sample with the CTM data, a georeferenced pixel is selected in each parcel within the two-stage sample.

To assess uncertainties, and particularly those due to classification errors, we use statistical models. The choice of a model depends on ground data, and we consider linear models, for ground data collected at the parcel level, which can be treated as values of a continuous variable, and multinomial logit models for ground data collected at the pixel level, which can be treated as values of a categorical variable.

3.1 Ground data collected at the parcel level

Let the sampling unit, i, be a parcel or a cluster of parcels (say a polygon, as in Spain), and let $i = 1, 2, \cdots, N$ be the set of sampling units of the population. We consider a linear model $y_i = \mathbf{x}_i\boldsymbol{\beta} + \varepsilon_i$ that relates ground data observed on the study crop in the ith sampling unit, y_i, with the CTM data, $\mathbf{x}_i$, aggregated at the ith level. For a given crop, $\mathbf{x}_i$ is a row vector with only two components; the first is the constant 1, $x_{1i} = 1$, and the second, $x_{2i} = x_i$, is the number of pixels classified in the CTM as belonging to the study crop in i: for convenience, we will denote $\mathbf{x}_i$ in the general form $\mathbf{x}_i = \underset{1 \le l \le L}{row}(x_{li})$ where $L = 2$, $x_{1i} = 1$ and $x_{2i} = x_i$. The parameters vector $\boldsymbol{\beta}$ is unknown and must be estimated.

In sampling units not included in the sample, only $\mathbf{x}_i$ is observed. Given $\mathbf{x}_i$, the expected value of the unknown y_i is modeled by $Ey_i = \mathbf{x}_i\boldsymbol{\beta}$, and the deviation of y_i from $\mathbf{x}_i\boldsymbol{\beta}$ is modeled by ε_i. This deviation is a source of uncertainty, mainly due to classification errors, and we will show how to assess it using the ε_i sampling variance.

The survey variable total in the population is $y_N = \sum_{i=1}^{N} y_i$. As a result of including the component $x_{1i} = 1$ in the model, y_N can be written as $y_N = \mathbf{x}_N \mathrm{B}_N$, where $\mathbf{x}_N = \sum_{i=1}^{N} \mathbf{x}_i = [N \ \ x_N]$, and $x_N = \sum_{i=1}^{N} x_i$ is the total number of pixels classified in the CTM as belonging to the study crop, and $\mathbf{B}_N = \left(\mathbf{X}_N^T \mathbf{X}_N\right)^{-1} \mathbf{X}_N^T \mathbf{y}_N$ designates the vector of population regression coefficients $\mathbf{B}_N = \underset{1 \le l \le L}{col}(B_l)$ (with $L = 2$), where $\mathbf{X}_N = \underset{1 \le i \le N}{col}\,\underset{1 \le l \le L}{row}(x_{li})$ (with $L = 2$) and $\mathbf{y}_N = \underset{1 \le i \le N}{col}(y_i)$ [11]. Since $\mathbf{x}_i$ is known for every population unit $i = 1, 2, \cdots, N$, then $\mathbf{x}_N$ and $\mathbf{X}_N$ are known. However, $\mathbf{y}_N$ is unknown and so is $\mathbf{B}_N$.

To estimate $\mathbf{B}_N$, we use the data collected in the sample n, selected from the population with inclusion probabilities $\{\pi_i; i = 1, 2, \cdots, N\}$. The design-based estimator $\hat{\mathbf{B}}_\pi = \left(\sum_{i=1}^{n} \frac{\mathbf{x}_i^T \mathbf{x}_i}{\pi_i}\right)^{-1} \sum_{i=1}^{n} \frac{\mathbf{x}_i^T y_i}{\pi_i}$ is design-consistent for $\mathbf{B}_N$, and as a result, the synthetic or projective estimator $\hat{y}_N = \mathbf{x}_N \hat{\mathbf{B}}_\pi$ is design-consistent for y_N [11].

The error of $\hat{y}_N$ as an estimator of y_N is $\hat{y}_N - y_N = \mathbf{x}_N\left(\hat{\mathbf{B}}_\pi - \mathbf{B}_N\right)$, and its sampling variance is $V\left(\hat{y}_N - y_N\right) = \mathbf{x}_N V\left(\hat{\mathbf{B}}_\pi - \mathbf{B}_N\right)\mathbf{x}_N^T$, where $V\left(\hat{\mathbf{B}}_\pi - \mathbf{B}_N\right) = \left(\sum_{i=1}^{N} \mathbf{x}_i^T \mathbf{x}_i\right)^{-1} V \sum_{i=1}^{n} \frac{\mathbf{x}_i^T \varepsilon_{iN}}{\pi_i} \left(\sum_{i=1}^{N} \mathbf{x}_i^T \mathbf{x}_i\right)^{-1}$ and $\varepsilon_{iN} = y_i - \mathbf{x}_i \mathbf{B}_N$. The sampling variance is a measure of uncertainty that depends both on the deviation ε_{iN} between the true value of y_i and its expected value based on $\mathbf{x}_i$ and on the inclusion probabilities, π_i. A design-consistent estimator of the sampling variance is $\hat{V}\left(\hat{y}_N - y_N\right) = V\left(\sum_{i=1}^{n} \frac{\hat{\varepsilon}_{iN}}{\pi_i}\right)$, where $V(.)$ is the design-based variance and $\hat{\varepsilon}_{iN} = y_i - \mathbf{x}_i \hat{\mathbf{B}}_\pi$.

The asymptotic distribution of $\hat{y}_N$ is normal, $\left[V\left(\hat{y}_N - y_N\right)\right]^{-1/2}\left(\hat{y}_N - y_N\right) \to N(0, 1)$, and can be used for constructing uncertainty measures in terms of probability for y_N, such as confidence intervals.

To estimate the total of the survey variable, $y_{N_R} = \sum_{i=1}^{N_R} y_i$, in large areas that are part of the population, such as a region or a province R with N_R sampling units within a country, we use the sample of size n selected from the population with inclusion probabilities $\{\pi_i; i = 1, 2, \cdots, N\}$ and the estimator $\hat{y}_{N_R} = \mathbf{x}_{N_R} \hat{\mathbf{B}}_\pi + \frac{N_R}{\hat{N}_R} \sum_{i=1}^{n_R} \frac{y_i - \mathbf{x}_i \hat{\mathbf{B}}_\pi}{\pi_i}$, where n_R is number of sampling units included in the sample n that fall in R, and $\hat{N}_R = \sum_{i=1}^{n} \frac{I_i}{\pi_i}$ is an estimator of N_R, where $I_i = 1$ if i is in R and $I_i = 0$ otherwise. A measure of the $\hat{y}_{N_R}$ uncertainty is its sampling variance $V\left(\hat{y}_{N_R} - y_{N_R}\right) = N_R^2 V \frac{1}{\hat{N}_R}\left(\sum_{i=1}^{n_R} \frac{\varepsilon_{iN}}{\pi_i}\right)$, and a design-based estimator of $V\left(\hat{y}_{N_R} - y_{N_R}\right)$ is $\hat{V}\left(\hat{y}_{N_R} - y_{N_R}\right) = N_R^2 V \frac{1}{\hat{N}_R}\left(\sum_{i=1}^{n_R} \frac{\hat{\varepsilon}_{iN}}{\pi_i}\right)$.

3.1.1 Crop type map relative efficiency

The relative efficiency of CTM data is $RE_{CTM} = V_G(V_{G+CTM})^{-1}$, where V_G is the current sampling variance based on ground data alone, and V_{G+CTM} is the sampling variance based on the integration of ground and CTM data. If $RE_{CTM} > 1$, then the current sampling variance may be reduced by the amount $V_G - V_{G+CTM} = \left(1 - RE_{CTM}^{-1}\right)V_G$, without increasing the sample size. As a result, the ground sample size can be reduced to $n_{G+CTM} = RE_{CTM}^{-1} n_G$ sampling units, without loss of accuracy. n_G is the current sample size for ground data collection.

For $\pi_i = n/N$, the sampling variance based on ground data alone is $V_G = V\sum_{i=1}^{n}\frac{y_i}{n/N} = N^2(1 - n/N)1/n(n-1)\sum_{i=1}^{n}(y_i - \bar{y})^2$ and the sampling variance using both ground and CTM data is $V_{G+CTM} = V\sum_{i=1}^{n}\frac{y_i - \mathbf{x}_i\hat{\mathbf{B}}_\pi}{n/N} = N^2(1 - n/N)1/n(n-2)\sum_{i=1}^{n}\left(y_i - \mathbf{x}_i\hat{\mathbf{B}}_\pi\right)^2$. As a result, $RE_{CTM} \simeq \sum_{i=1}^{n}(y_i - \bar{y})^2\left(\sum_{i=1}^{n}\left(y_i - \mathbf{x}_i\hat{\mathbf{B}}_\pi\right)^2\right)^{-1}$ and $n_{G+CTM} = \left(\sum_{i=1}^{n}\left(y_i - \mathbf{x}_i\hat{\mathbf{B}}_\pi\right)^2\right)\left(\sum_{i=1}^{n}(y_i - \bar{y})^2\right)^{-1} n_G$.

3.1.2 Example

Our data source is Sen4Stat. The goal of this project was focused on developing an open-source system for generating CTM from Sentinel images, using data provided by the NSOs (https://www.esa-sen4stat.org/). We use field data on crops collected at the parcel level by the NSO of Spain, to illustrate the integration of continuous ground data and CTM data using linear models.

3.1.2.1 Crop acreage estimates

We consider a large area of size 100 × 100 km in Castilla y León (Spain), with X coordinates (in meters): (300,000,400,000) and Y coordinates (4,600,000,4,700,000). The sampling unit is a square with a side of 700 m, and the sample size is 419 sampling units. The study crop is barley, and the ground data were observed at the parcel level. However, data were aggregated at the sampling unit level for calculations so that y_i represents the ground data on the barley acreage in sampling unit i and in $\mathbf{x}_i = [1 \;\; x_i]$, x_i is the number of pixels classified in CTM as belonging to barley in this same sampling unit i.

Model parameter estimates are $\hat{\mathbf{B}}_\pi = [0.046 \;\; 0.093]^T$, and barley acreage estimates based both on ground data alone and on the integration of ground and CTM data are found in **Table 1**. As a result of this integration, the accuracy of the estimates improves considerably: the sampling error (or coefficient of variation, which is the ratio between the square root of the sampling variance and the acreage estimate in %) and the wide of the confidence interval (within whose limits the true value of the barley acreage lies, with a high probability (95%) were reduced by half.

The CTM relative efficiency, RE_{CTM}, was high: using CTM data, the current ground sample size could be reduced to less than one-fifth without loss of accuracy.

Data	Acreage (hectare)	Uncertainty			Relative efficiency of CTM data
		95% Confidence interval (hectare)		Sampling error (CV%)	
		Limits	Wide		
Ground	236165.4	Lw:15951.7	40427.3	4.37	—
		Up: 56379			
Ground + CTM	228550.1	Lw:19699.8	17700.5	1.98	5.2
		Up:37400.3			

Table 1.
Barley acreage estimates in a 100 × 100 km area of Castilla y León, 2018.

3.1.2.2 Crop yield estimates

Let y_{ij} be the yield of barley observed in parcel j of sampling unit i, and let $\mathbf{x}_{ij} = \underset{1 \le l \le L}{row}\left(x_{ijl}\right)$ be a $1 \times L$ row vector with $L = 2$, $x_{ij1} = 1$ in the first position, and a normalized difference vegetation index (NDVI) in the second position, . The estimates based both on ground data alone and on the integration of ground and NDVI data are found in **Table 2**.

Although the vegetation index contribution to explain crop yield is statistically significant, its contribution to improve yield estimator accuracy is low: the sampling error is of the same order of magnitude as if ground data alone were used. The relative efficiency is nearly 1, so the current sample size could barely be reduced without loss of accuracy. Additional research is required to make production estimates more reliable through improved yield estimation from RS data.

3.1.2.3 Crop production estimates

Table 3 shows the estimate production of barley, calculated as the product of the crop acreage estimate and yield estimate.

Using RS data, the production estimator accuracy increased considerably: the estimation error decreased by half, even if the RS data is not reliable for crop yield.

Data	Yield (kg/hectare)	Uncertainty			Relative efficiency of RS data
		95% Confidence interval (kg/hectare)		Sampling error (CV%)	
		Limits	Wide		
Ground	4213.617	Lw: 4093.97	239.293	1.45	—
		Up: 4333.263			
Ground + RS	4155.688	Lw: 4033.231	244.915	1.50	0.95
		Up: 4278.146			

Table 2.
Barley yield estimates in a 100 × 100 km area of Castilla y León, 2018. NDVI.

Data	Production (tons) (1000 kg)	Uncertainty		
		95% Confidence interval (tons: 1000 kg)		Sampling error (CV%)
		Limits	Wide	
Ground	987,841	Lw: 903640	168,402	4.35
		Up: 1072042		
Ground + RS	965,733	Lw: 915194	101,078	2.67
		Up: 1016272		

Table 3.
Barley production estimates in a 100 × 100 km area of Castilla y León, 2018.

3.1.2.4 Crop acreage estimates at the provincial level

We estimate barley acreage in that part of the provinces within the 100 × 100 km area of Castilla y León (Spain). Results are in **Table 4**.

The estimate accuracy at provincial level is less than that at the national level, but the CTM relative efficiency is similar. Where the sampling variance using ground data alone is high (León), its integration with CTM data reduces the sampling error below the standard limits (20%) considered acceptable in official statistics.

3.2 Ground data collected at the pixel level

We use pixel i as sampling unit, in the same way that a point is used, that is, treating the ground data observed at the pixel level as values of a categorical variable. To each pixel $i = 1, 2, \cdots, N$ of the study area, we associate a survey vector $\mathbf{y}_i = \underset{1 \le k \le K}{col}(y_{ik})$, where $y_{ik} = 1$ if crop k covers pixel i and $y_{ik} = 0$ otherwise. K is the total number of crops so that the constraint $\sum_{k=1}^{K} y_{ik} = 1$ holds.

The total of this survey vector in the population is $\mathbf{y}_N = \sum_{i=1}^{N} \mathbf{y}_i = \underset{1 \le k \le K}{col}\left(\sum_{i=1}^{N} y_{ik}\right) = \underset{1 \le k \le K}{col}(y_{Nk})$, where y_{Nk} is the total number of pixels covered by crop k. It is assumed that a pixel is covered by a single crop (or that a class of mixed pixels is included) so that the area covered by k is $A_k = a y_{Nk}$, where a is the area of the terrain represented by a pixel.

Province	Ground data alone		Ground & CTM data		Relative efficiency of CTM data
	Acreage (has.)	Sampling error (CV%)	Acreage (has.)	Sampling error (CV%)	
León	6853.2	24.15	6834.5	16.20	2.3
Palencia	88602.0	7.36	90535.3	3.33	4.7
Valladolid	128209.5	5.57	119707.4	2.66	5.1
Zamora	12324.2	17.71	10948.2	8.16	6.4
Total area	235989.1	4.37	228028.5	1.98	5.2

Table 4.
Barley acreage estimates at provincial level.

To estimate $\mathbf{y}_N$, we follow a design-based approach using ground data collected in a sample of n pixels, selected from the population with inclusion probabilities $\{\pi_i; i = 1, 2, \cdots, N\}$. To integrate these sample data with CTM data, we use a multinomial logit model [12–14]. We assume that the survey vector $\mathbf{y}_i = \underset{1 \le k \le K}{col}(y_{ik})$ follows a multinomial distribution $MN(1, \boldsymbol{\mu}_i)$, where $\boldsymbol{\mu}_i = \underset{1 \le k \le K}{col}(\mu_{ik})$ and μ_{ik} is the probability that crop $k = 1, 2, \cdots, K$ covers pixel i (probability of $y_{ik} = 1$ and $y_{ik'} = 0; \forall k' \ne k$), with the constraint $\sum_{k=1}^{K} \mu_{ik} = 1$.

To link μ_{ik} with CTM data, we use a logit model $\mu_{ik} = \frac{\exp(\mathbf{x}_i \boldsymbol{\beta}_k)}{\sum_{k=1}^{K} \exp(\mathbf{x}_i \boldsymbol{\beta}_k)}$, where $\mathbf{x}_i = \underset{1 \le l \le L}{row}(x_{il})$ is an indicator vector of the class where pixel i is located within the CTM ($x_{il} = 1$ if pixel i is in the class l and $x_{il} = 0$ otherwise), and $\boldsymbol{\beta}_k = \underset{1 \le l \le L}{col}(\beta_{kl})$ is an unknown parameter vector. To estimate μ_{ik}, it is sufficient to obtain estimates of $\boldsymbol{\beta}_k$ for $k = 1, 2, \cdots, K - 1$ [12]. The probability of crop K is $\mu_K = 1 - \sum_{k=1}^{K-1} \mu_{ik}$ and can be estimated using the estimates $\hat{\boldsymbol{\beta}}_{\pi k}$ for $k = 1, 2, \cdots, K - 1$.

The design-based parameter estimator $\hat{\mathbf{B}}_\pi = \underset{1 \le k \le K-1}{col}(\hat{\mathbf{B}}_{\pi k}) = \underset{1 \le k \le K-1}{col}\ \underset{1 \le l \le L}{col}(\hat{B}_{\pi kl})$ is design-consistent and can be found iteratively using $\hat{\mathbf{B}}_\pi^{(m+1)} = \hat{\mathbf{B}}_\pi^{(m)} +$

$$\left[\sum_{i=1}^{n} \frac{\mathbf{H}(y_i, \boldsymbol{\beta})}{\pi_i} \bigg|_{\mathbf{B} = \hat{\mathbf{B}}_\pi^{(m)}} \right]^{-1} \sum_{i=1}^{n} \frac{\mathbf{b}(y_i, \boldsymbol{\beta})}{\pi_i} \bigg|_{\mathbf{B} = \hat{\mathbf{B}}_\pi^{(m)}}$$

[11], where $\mathbf{b}(y_i, \boldsymbol{\beta}) = \underset{1 \le k \le K-1}{col}\ \underset{1 \le l \le L}{col}\left[(y_{ik} - \mu_{ik}) x_{il}\right]$ and $\mathbf{H}(y_i, \boldsymbol{\beta}) = \left[\underset{1 \le k \le K-1}{col}\ \underset{1 \le k' \le K-1}{row}(\delta_{kk'} \mu_{ik} - \mu_{ik} \mu_{ik'}) \right] \otimes \mathbf{x}_i^T \mathbf{x}_i$, with $\delta_{kk'} = 1$ if $k = k'$ and $\delta_{kk'} = 0$ otherwise.

A design-consistent estimator of $\mathbf{y}_N = \underset{1 \le k \le K}{col}(y_{Nk})$ is $\hat{\mathbf{y}}_N = \sum_{i=1}^{N} \hat{\boldsymbol{\mu}}_i + \frac{N}{\hat{N}_p} \sum_{i=1}^{n} \frac{\mathbf{y}_i - \hat{\boldsymbol{\mu}}_i}{\pi_i} = \underset{1 \le k \le K-1}{col}(\hat{y}_{Nk})$, where $\hat{\boldsymbol{\mu}}_i = \underset{1 \le k \le K-1}{col}(\hat{\mu}_{ik})$, $\hat{\mu}_{ik} = \frac{\exp(\mathbf{x}_i \hat{\mathbf{B}}_{\pi k})}{1 + \sum_{k=1}^{K-1} \exp(\mathbf{x}_i \hat{\mathbf{B}}_{\pi k})}$ for $k = 1, 2, \cdots, K - 1$, $\hat{\mu}_{iK} = 1 - \sum_{k=1}^{K-1} \hat{\mu}_{ik} = \frac{1}{1 + \sum_{k=1}^{K-1} \exp(\mathbf{x}_i \hat{\mathbf{B}}_{\pi k})}$, $\hat{N}_p = \sum_{i=1}^{n} \frac{1}{\pi_i}$, and $\hat{y}_{Nk} = \sum_{i=1}^{N} \hat{\mu}_{ki} + \frac{N}{\hat{N}_p} \sum_{i=1}^{n} \frac{y_{ki} - \hat{\mu}_{ki}}{\pi_i}$ is the estimator of the total number of pixels covered by crop k for $k = 1, 2, \cdots, K - 1$; for crop K, it is $\hat{y}_{NK} = N - \sum_{k=1}^{K-1} \hat{y}_{Nk}$.

The sampling covariance matrix of $\hat{\mathbf{y}}_N$ is $\mathbf{V}\hat{\mathbf{y}}_N = N^2 \mathbf{V} \frac{1}{\hat{N}_p} \sum_{i=1}^{n} \frac{\mathbf{y}_i - \hat{\boldsymbol{\mu}}_i}{\pi_i} = \underset{1 \le k \le K-1}{col}\ \underset{1 \le k' \le K-1}{row}\left(N^2 Cov(\hat{y}_{rk}, \hat{y}_{rk'})\right)$, where $\hat{y}_{rk} = \frac{1}{\hat{N}_p} \sum_{i=1}^{n} \frac{y_{ki} - \mu_{ki}}{\pi_i}$ is a function of $\hat{N}_{pk} = \sum_{i=1}^{n} \frac{y_{ki}}{\pi_i}$ (because $\hat{N}_p = \sum_{k=1}^{K} \hat{N}_{pk}$), the estimators of the total number of pixels covered by crop $k = 1, 2, \cdots, K$ based on ground data alone, and of $\hat{r}_{kN} = \sum_{i=1}^{n} \frac{y_{ki} - \hat{\mu}_{ki}}{\pi_i}$, the estimator of the total of the deviations between y_{ki} and μ_{ki} in the population.

The sampling variance is an uncertainty measure that depends both on the deviation between the true, $\mathbf{y}_i$, value and its expected value estimate, $\hat{\boldsymbol{\mu}}_i$, and on the sample design, π_i. We focus on the sampling variance $V\hat{y}_{Nk}$ for $k = 1, 2, \cdots, K - 1$, which are the elements along the diagonal of $\mathbf{V}\hat{\mathbf{y}}_N$. Let $\hat{\mathbf{G}}_k = \underset{1\le j\le K+1}{row}\left(g_j\right) = \left[\hat{r}_{kN}\ \hat{N}_{p1}\ \hat{N}_{p2}\cdots\ \hat{N}_{pK}\right]$ be a row vector whose $K+1$ components are the estimators on which $\hat{y}_{rk}$ depends. The $\hat{y}_{Nk}$ sampling variance is $V\hat{y}_{Nk} = \underset{1\le j\le K+1}{row}\left(\frac{\partial \hat{y}_{rk}}{\partial g_j}\right)\mathbf{V}\hat{\mathbf{G}}_k \underset{1\le j\le K+1}{col}\left(\frac{\partial \hat{y}_{rk}}{\partial g_j}\right)$, where $\mathbf{V}\hat{\mathbf{G}}_k = \underset{1\le j\le K+1}{col}\ \underset{1\le j'\le K+1}{row}\left(Cov\left(g_j, g_{j'}\right)\right)$ is the design-based covariance matrix of $\hat{\mathbf{G}}_k$. The sampling covariance matrix of $\hat{\mathbf{y}}_N$ is estimated by replacing $\boldsymbol{\mu}_i$ with $\hat{\boldsymbol{\mu}}_i$ in$\mathbf{V}\hat{\mathbf{y}}_N$.

3.2.1 Crop type map relative efficiency

The sampling variance of the y_{Nk} estimator based on ground data alone is $V_G = N^2 V \frac{1}{\hat{N}_p}\sum_{i=1}^{n}\frac{y_{ki}}{\pi_i}$ and that based on its integration with CTM data is $V_{G+CTM} = N^2 V \frac{1}{\hat{N}_p}\sum_{i=1}^{n}\frac{y_{ki}-\mu_{ki}}{\pi_i}$. The relative efficiency of CTM data for estimating the total number of pixels covered by crop k is $RE_{CTMk} = V\frac{1}{\hat{N}_p}\sum_{i=1}^{n}\frac{y_{ki}}{\pi_i} \times \left(V\frac{1}{\hat{N}_p}\sum_{i=1}^{n}\frac{y_{ki}-\mu_{ki}}{\pi_i}\right)^{-1}$.

3.2.2 Example: crop acreage estimates

We use field data collected at the pixel level by the NSO of Senegal, to illustrate the integration of categorical ground data and CTM data using multinomial logit models. The crop covering each pixel included in the sample selected in the Nioro Department was observed in 2021. In the CTM, pixels covering agricultural areas were classified into four crop types: (i) maize, (ii) millet, (iii) groundnut, and (iv) other crops, and the data are encode as $\mathbf{x}_i = [1\ \ 0\ \ 0\ \ 0]$ for any pixel i of the class maize, $\mathbf{x}_i = [0\ \ 1\ \ 0\ \ 0]$ for any pixel i of the class millet, $\mathbf{x}_i = [0\ \ 0\ \ 1\ \ 0]$ for any pixel i of the class groundnut, and $\mathbf{x}_i = [0\ \ 0\ \ 0\ \ 1]$ for any pixel i of the class other crops.

We focus on estimating the acreage of the two main crops: millet and groundnut. The remaining crops (mainly maize) are included in a third category called other, whose estimated probability is one minus the estimated probability of millet and groundnut. Model parameter estimates are in **Table 5**.

Crop acreage estimates of millet and groundnut based on ground and CTM data are in **Table 6**.

The CTM data efficiency estimate is in **Table 7**.

The integration of ground data observed at the pixel level and CTM data improves the accuracy of the estimators based on ground data alone: the standard error is cut by a half, without loss of design-based consistency. As a result, the size of the current sample used for collecting ground data could be reduced to less than a third, without loss of accuracy.

Crop	Crop type map			
	Maize	Millet	Groundnut	Other crops
Millet	−0.010655	1.47958334	0.88931346	8.76504093
Groundnut	−0.659204	−0.59871700	2.89873948	1.21898514

Table 5.
Model parameter estimates ($\hat{\mathrm{B}}_{\pi millet}$ and $\hat{\mathrm{B}}_{\pi groundnut}$).

Crop	Acreage (Hectare)	Uncertainty			
		Sampling error (CV%)	Limits of 95% Confidence interval		
			Lower	Upper	Wide
Millet	89,215	4.11	81978.88	96330.4	14351.52
Groundnut	78,815	3.71	73089.15	84550.98	11461.82

Table 6.
Crop acreage estimates, Nioro (Senegal) 2021.

Crop type	Standard errors of proportion estimators		Relative efficiency of CTM data
	Ground data alone	Ground and CTM data	
Millet	3.37	1.90	3.13
Groundnut	3.34	1.52	4.80

Table 7.
Efficiency of CTM data for crop acreage estimation, Nioro (Senegal).

District	Millet		Groundnut	
	Acreage (Has.)	Sampling error (CV%)	Acreage (has.)	Sampling error (CV%)
Medina Sabakh	20067.21	8.6	19765.36	7.3
Paoskoto	38316.02	5.3	35018.23	4.0
Wack Ngouna	30831.77	11.9	24030.71	10.7
Total Nioro	89215.00	4.11	78815.00	3.71

Table 8.
Crop acreage estimates at the district level, Nioro (Senegal).

Table 8 shows crop acreage estimates in the districts Medina Sabakh, Paoskoto, and Wack Ngouna.

The estimate accuracy at the district level is lower than at the department level. However, the sampling error is within the standard limit ($CV < 20\%$) for labeling as official statistics.

4. Model-based approach: small area estimation

In small areas, such a municipality, the sample size is small, and as a result, the design-based approach is not accurate enough for most applications. In these areas, inference is based on the model, rather than the sampling design, and the estimates are not robust, but model-dependent.

4.1 Ground data collected at the parcel level

Let the sampling unit, i, be a parcel or a cluster of parcels (say a polygon, as in Spain), and let y_{di} be the value of the survey variable in a sampling unit $i = 1, 2, \cdots, N_d$ of the small area $d = 1, 2, \cdots, D$. Let $\left\{ \left(y_{di}, \mathbf{x}_{di} \right), i = 1, 2, \cdots, n_d \right\}$ be the ground data set,

y_{di}, and the CTM data set, $\mathbf{x}_{di} = [1 \ x_{d2i}]$, in the n_d sampling units from the whole sample n that fall in d. Here, x_{d2i} is the number of pixels classified in CTM as belonging to the study crop within the sampling unit i of the small area d.

To estimate $y_{N_d} = \sum_{i=1}^{N_d} y_{di}$, the total of the survey variable in d, we use CTM data and the linear mixed model $y_{di} = \mathbf{x}_{di}\boldsymbol{\beta} + u_d + \varepsilon_{di}$, where (u_d, ε_{di}) are zero-mean independent random variables of variances (σ_u^2, σ_e^2).

The model-based estimator of y_{N_d} is

$\hat{y}_{N_d} = N_d\left[(1-\hat{\gamma}_d)\overline{\mathbf{x}}_{N_d}\hat{\boldsymbol{\beta}} + \hat{\gamma}_d\left(\overline{y}_{n_d} + (\overline{\mathbf{x}}_{N_d} - \overline{\mathbf{x}}_{n_d})\hat{\boldsymbol{\beta}}\right)\right]$, where $\hat{\boldsymbol{\beta}} = \left(\mathbf{X}^T\hat{\mathbf{V}}^{-1}\mathbf{X}\right)^{-1}\mathbf{X}^T\hat{\mathbf{V}}^{-1}\mathbf{y}$ is the generalized least-square estimator of $\boldsymbol{\beta}$, with $\mathbf{y} = \underset{1\le d\le D}{col}\ \underset{1\le i\le n_d}{col}(y_{di})$,

$\mathbf{X} = \underset{1\le d\le D}{col}\ \underset{1\le i\le n_d}{col}(\mathbf{x}_{di})$, $\hat{\mathbf{V}}^{-1} = \underset{1\le d\le D}{diag}\left(\hat{\mathbf{V}}_d^{-1}\right)$ and $\hat{\mathbf{V}}_d^{-1} = \frac{1}{\hat{\sigma}_e^2}\mathbf{I}_{n_d} - \frac{\hat{\gamma}_d}{n_d\hat{\sigma}_e^2}\mathbf{1}_{n_d}\mathbf{1}_{n_d}^T$, where

$\hat{\gamma}_d = \frac{\hat{\sigma}_u^2}{\hat{\sigma}_u^2 + \hat{\sigma}_e^2/n_d}$. Here, $\hat{\sigma}_e^2 = \frac{\mathbf{e}^T\mathbf{e}}{n-D-1}$ and $\hat{\sigma}_u^2 = \frac{\mathbf{u}^T\mathbf{u} - (n-2)\hat{\sigma}_e^2}{n_*}$, with

$n_* = \sum_{d=1}^{D} n_d\left[1 - n_d\overline{\mathbf{x}}_d\left(\mathbf{X}^T\mathbf{X}\right)^{-1}\overline{\mathbf{x}}_d^T\right]$, are unbiased estimators of the variance components, where $\mathbf{e}^T\mathbf{e}$ is the sum of squared residuals in the model fitted by ordinary least squares, taking as fixed the small-area effect u_d. $\mathbf{u}^T\mathbf{u}$ is the sum of squared residuals in the model fitted by ordinary least squares, with $u_d = 0$. $\overline{\mathbf{x}}_{N_d}$ and $\overline{\mathbf{x}}_{n_d}$ are the population and sample means of the vector $\mathbf{x}_{di}$, respectively.

An unbiased estimator of the total mean-square error estimator $MSE\left(\hat{y}_{N_d}\right)$ is [15]:

$\hat{MSE}\left(\hat{y}_{N_d}\right) = N_d^2\left(1 - \frac{n_d}{N_d}\right)^2\left[h_{1d}(\hat{\sigma}_v^2, \hat{\sigma}_e^2) + h_{2d}(\hat{\sigma}_v^2, \hat{\sigma}_e^2) + 2\,h_{3d}(\hat{\sigma}_v^2, \hat{\sigma}_e^2)\right]$, where:

$$h_{1d}(\hat{\sigma}_v^2, \hat{\sigma}_e^2) = \hat{\gamma}_d\left(\frac{\hat{\sigma}_e^2}{n_d}\right) + \left(1 - \frac{n_d}{N_d}\right)^2\frac{(N_d - n_d)}{N_d^2}$$

$$h_{2d}(\hat{\sigma}_v^2, \hat{\sigma}_e^2) = \hat{\sigma}_e^2\left(\overline{\mathbf{X}}_{N_d-n_d} - \hat{\gamma}_d\,\overline{\mathbf{x}}_d\right)\mathbf{A}^{-1}\left(\overline{\mathbf{X}}_{N_d-n_d} - \hat{\gamma}_d\,\overline{\mathbf{x}}_d\right)^T$$

$$h_{3d}(\hat{\sigma}_v^2, \hat{\sigma}_e^2) = \frac{1}{n_d^2}\frac{1}{\left(\hat{\sigma}_v^2 + \frac{\hat{\sigma}_e^2}{n_d}\right)^3}\left[\left(\hat{\sigma}_e^2\right)^2 V\hat{\sigma}_v^2 + \left(\hat{\sigma}_v^2\right)^2 V\hat{\sigma}_e^2 - 2\hat{\sigma}_e^2\ \hat{\sigma}_v^2\ Cov(\hat{\sigma}_e^2, \hat{\sigma}_v^2)\right]$$

(where)

$$\mathbf{A} = \sum_{d=1}^{D}\left(\sum_{i=1}^{n_d}\mathbf{x}_{di}^T\mathbf{x}_{di} - \hat{\gamma}_d n_d\overline{\mathbf{x}}_{n_d}^T\overline{\mathbf{x}}_{n_d}\right)$$

$$V(\hat{\sigma}_v^2) = \frac{2}{n_*^2}\left[\frac{1}{n-D-1}(D-1)(n-2)\left(\hat{\sigma}_e^2\right)^2 + 2n_*\hat{\sigma}_e^2\ \hat{\sigma}_v^2 + n_{**}\left(\hat{\sigma}_v^2\right)^2\right]$$

$$V(\hat{\sigma}_e^2) = \frac{2\left(\hat{\sigma}_e^2\right)^2}{n-D-1}$$

$$Cov(\hat{\sigma}_e^2, \hat{\sigma}_v^2) = -\frac{1}{n_*}(D-1)V\hat{\sigma}_e^2$$

Here, $n_{**} = \sum_{d=1}^{D} n_d^2\left(1 - n_d\overline{\mathbf{x}}_{n_d}\mathbf{A}_1^{-1}\overline{\mathbf{x}}_{n_d}^T\right) + tr\left(\mathbf{A}_1^{-1}\sum_{d=1}^{D} n_d^2\overline{\mathbf{x}}_{n_d}^T\overline{\mathbf{x}}_{n_d}\right)^2$. Because $\mathbf{A}_1 = \sum_{d=1}^{D}\sum_{i=1}^{n_d}\mathbf{x}_{di}^T\mathbf{x}_{di}$, n_{**} may be simplified to $n_{**} = \sum_{d=1}^{D} n_d^2\left[1 - \overline{\mathbf{x}}_{n_d}\left(\mathbf{X}^T\mathbf{X}\right)^{-1}\overline{\mathbf{x}}_{n_d}^T\right] = n_* - n + \sum_{d=1}^{D} n_d^2$.

4.1.1 Example

Table 9 shows barley acreage estimates at the municipality level in that part of Zamora province within the 100 × 100 km area of Castilla y León (Spain).

Even in municipalities where the sample size is as small as a single sampling unit, CTM data contribute to reducing sampling error below the standard limit (20%) considered to label estimates as official statistics.

4.2 Ground data collected at the pixel level

We consider a population of $i = 1, 2, \cdots, N$ pixels grouped in a set of $d = 1, 2, \cdots, D$ small areas (municipalities) such that $N = \sum_{d=1}^{D} N_d$, where N_d is the number of pixels in the small area d. Let $\mathbf{y}_{di} = \underset{1 \le k \le K}{col}(y_{dik})$ be the survey vector where $y_{dik} = 1$ when crop k covers pixel i and $y_{dik} = 0$ otherwise.

To estimate $\mathbf{y}_{N_d} = \sum_{i=1}^{N_d} \mathbf{y}_{di} = \underset{1 \le k \le K}{col}\left(y_{N_d k}\right)$, the total of the survey vector over a small area d, where $y_{N_d k} = \sum_{i=1}^{N_d} y_{dik}$ is the number of pixels of d covered by crop k, we use CTM data, and we assume that $\mathbf{y}_{di}$ follows a multinomial distribution $MN(1, \boldsymbol{\mu}_{di})$, where $\boldsymbol{\mu}_{di} = \underset{1 \le k \le K}{col}(\mu_{dik})$ and μ_{dik} is the probability of $y_{dik} = 1$ and $y_{dik'} = 0; \forall k' \neq k$, with the constraint $\sum_{k=1}^{K} \mu_{dik} = 1$. To link μ_{dik} with CTM data, we use a logit mixed model $\mu_{dik} = \frac{\exp(\mathbf{x}_{di}\boldsymbol{\beta}_k + v_{dk})}{1 + \sum_{k=1}^{K-1} \exp(\mathbf{x}_{di}\boldsymbol{\beta}_k + v_{dk})}$, where μ_{dik} is the probability that crop k covers i of d, $\mathbf{x}_{di} = \underset{1 \le l \le L}{row}(x_{dil})$ is an indicator vector whose components are $x_{dil} = 1$ if pixel i of d is classified in the crop-type class $l = 1, 2, \cdots, L$ and $x_{dil} = 0$ otherwise, and $\mathbf{x}_{di}\boldsymbol{\beta}_k + v_{dk} =$

Municipality	Sample size	Estimates	
		Acreage (Hectares)	Sampling error (CV%)
Belver de los Montes	1	212.96	29.1
Castroverde	3	2914.22	8.0
Pinilla de Toro	4	963.30	10.0
Quintanilla del Monte	1	466.65	20.3
Toro	1	615.91	14.0
Vezdemarbán	3	1358.22	12.6
Villalpando	2	560.05	39.1
Villamayor de Campos	1	1056.23	11.1
Villanueva del Campo	1	784.03	13.2
Villar de Fallaves	1	844.16	11.0
Villardondiego	1	516.40	11.5
Villavendimio	1	656.07	10.4
Total Zamora	20	10948.2	8.16

Table 9.
Barley acreage estimates at municipality level.

η_{dik} is a linear mixed predictor consisting of a fixed $\mathbf{x}_{di}\boldsymbol{\beta}_k$ and a random component, v_{dk}, associated with the small areas and specifics for each crop k.

Here, $\boldsymbol{\beta}_k = \underset{1\le l\le L}{col}(\beta_{kl})$ is a vector of parameters measuring the effects of the CTM predictors $\mathbf{x}_{di}$ on $\boldsymbol{\mu}_{di}$, for example, for a pixel of the crop-type class $l = 1, 2, \cdots, L$ we have $\mu_{dik} = \frac{\exp(\beta_{kl}+v_{dk})}{1+\sum_{k=1}^{K-1}\exp(\beta_{kl}+v_{dk})}$ for crops $k = 1, 2, \cdots, K-1$ and $\mu_{diK} = \frac{1}{1+\sum_{k=1}^{K-1}\exp(\beta_{kl}+v_{dk})}$ for crop K. The random effect, v_{dk}, accounts for the heterogeneity of the survey vector $\mathbf{y}_{di} = \underset{1\le k\le K}{col}(y_{dik})$ among small areas, not explained by CTM data. It is assumed that v_{dk} is a realization of a zero-mean random variable whose small-area variance is $Vv_{dk} = \sigma_k^2$ for crop k and whose cross-crop covariance is $Cov(v_{dk}, v_{dk'}) = \sigma_{kk'}^2$ so that the vector of small-area random effects $\mathbf{V} = \underset{1\le d\le D}{col}\underset{1\le k\le K-1}{col}(v_{dk})$ has mean $E\mathbf{V} = \mathbf{0}$ and covariance matrix $\boldsymbol{\Sigma} = \boldsymbol{\theta}\otimes\mathbf{I}_D$, where $\boldsymbol{\theta} = \underset{1\le k\le K-1}{row}\underset{1\le k'\le K-1}{col}\left(\sigma_{kk'}^2\right)$ and $\sigma_{kk'}^2 = \sigma_k^2$ if $k = k'$.

To estimate $\mathbf{y}_{N_d}$, we observe $\mathbf{y}_{di}$ only in the n_d pixels that fall within the small area d, that is, a small part of the full sample of size $n = \sum_{d=1}^{D} n_d$; the values of $\mathbf{y}_{di}$ in the remaining $N_d - n_d$ pixels are unknown, and we want to predict them using the model. We consider $\mathbf{y}_{N_d}$ decomposed into two $\mathbf{y}_{di}$ aggregates $\mathbf{y}_{N_d} = \sum_{i_s=1}^{n_d}\mathbf{y}_{di_s} + \sum_{i_r=1}^{N_d-n_d}\mathbf{y}_{di_r}$: one known, $\sum_{i_s=1}^{n_d}\mathbf{y}_{di_s}$, and the other, $\sum_{i_r=1}^{N_d-n_d}\mathbf{y}_{di_r}$, unknown. We use $\hat{\mathbf{y}}_{N_d} = \sum_{i_s=1}^{n_d}\mathbf{y}_{di_s} + \sum_{i_r=1}^{N_d-n_d}\hat{\boldsymbol{\mu}}_{di_r}$ as a predictor of $\mathbf{y}_{N_d}$, where $\hat{\boldsymbol{\mu}}_{di_r} = \underset{1\le k\le K-1}{col}\left(\hat{\mu}_{di_rk}\right)$ with $\hat{\mu}_{di_rk} = \frac{\exp\left(\mathbf{x}_{di_r}\hat{\boldsymbol{\beta}}_k+\hat{v}_{dk}\right)}{1+\sum_{k=1}^{K-1}\exp\left(\mathbf{x}_{di_r}\hat{\boldsymbol{\beta}}_k+\hat{v}_{dk}\right)}$ for crop $k = 1, 2, \cdots, K-1$, and $\hat{\mu}_{di_rK} = \frac{1}{1+\sum_{k=1}^{K-1}\exp\left(\mathbf{x}_{di_r}\hat{\boldsymbol{\beta}}_k+\hat{v}_{dk}\right)}$ for crop K. Here, $\hat{\boldsymbol{\beta}}_k$ is the $\boldsymbol{\beta}_k$ estimate and $\hat{v}_{dk}$ is the predictor of the small-area random effect v_{dk} in $\mathbf{v} = \underset{1\le d\le D}{col}\underset{1\le k\le K-1}{col}(v_{dk})$. To estimate both the fixed-effect $\boldsymbol{\beta} = \underset{1\le k\le K-1}{col}(\boldsymbol{\beta}_k)$ and random-effect $\boldsymbol{\theta} = \underset{1\le k\le K-1}{row}\underset{1\le k'\le K-1}{col}\left(\sigma_{kk'}^2\right)$ model parameters, we use maximum-likelihood estimators and the big sample of size $n = \sum_{d=1}^{D} n_d$. As a result, both predictors $\hat{\boldsymbol{\mu}}_{di_r}$ and $\hat{\mathbf{y}}_{N_d}$ are asymptotically unbiased.

The mean squared error of the predictor $\hat{\mathbf{y}}_{N_d}$ of $\mathbf{y}_{N_d}$ is $MSE\left(\hat{\mathbf{y}}_{N_d}\right) = E\left(\hat{\mathbf{y}}_{N_d} - \mathbf{y}_{N_d}\right)\left(\hat{\mathbf{y}}_{N_d} - \mathbf{y}_{N_d}\right)^T = E\left[\sum_{i_r=1}^{N_d-n_d}\left(\hat{\boldsymbol{\mu}}_{di_r} - \mathbf{y}_{di_r}\right)\right]^2$, where $\sum_{i_r=1}^{N_d-n_d}\left(\hat{\boldsymbol{\mu}}_{di_r} - \mathbf{y}_{di_r}\right) = \sum_{i_r=1}^{N_d-n_d}\left(\hat{\boldsymbol{\mu}}_{di_r} - \boldsymbol{\mu}_{di_r}\right)\left(\mathbf{y}_{di_r} - \boldsymbol{\mu}_{di_r}\right) = \hat{\boldsymbol{\tau}}_{dr} - \boldsymbol{\tau}_{dr} - \boldsymbol{\varepsilon}_{dr}$. Here, $\boldsymbol{\tau}_{dr} = \sum_{i_r=1}^{N_d-n_d} E\mathbf{y}_{di_r} = \sum_{i_r=1}^{N_d-n_d}\hat{\boldsymbol{\mu}}_{di_r}$ is the expected value of the survey vector in the $N_d - n_d$ non-observed pixels, $\hat{\boldsymbol{\tau}}_{dr} = \sum_{i_r=1}^{N_d-n_d}\hat{\boldsymbol{\mu}}_{di_r}$ is the predictor of $\boldsymbol{\tau}_{dr}$, and $\boldsymbol{\varepsilon}_{dr} = \sum_{i_r=1}^{N_d-n_d}\boldsymbol{\varepsilon}_{di_r}$ with $\boldsymbol{\varepsilon}_{di_r} = \left(\mathbf{y}_{di_r} - \boldsymbol{\mu}_{di_r}\right)$ is the unobservable error vector.

Then, $MSE\left(\hat{\mathbf{y}}_{N_d}\right) = E[\hat{\boldsymbol{\tau}}_{dtr} - \boldsymbol{\tau}_{dtr} - \boldsymbol{\varepsilon}_{dtr}]^2 = MSE(\hat{\boldsymbol{\tau}}_{dr}) + E\boldsymbol{\varepsilon}_{dr}\boldsymbol{\varepsilon}_{dr}^T - E(\hat{\boldsymbol{\tau}}_{dr} - \boldsymbol{\tau}_{dr})\boldsymbol{\varepsilon}_{dr}^T - E\boldsymbol{\varepsilon}_{dr}(\hat{\boldsymbol{\tau}}_{dr} - \boldsymbol{\tau}_{dr})^T$. If we assumed that $\boldsymbol{\varepsilon}_{dr}$ and $\hat{\boldsymbol{\tau}}_{dr}$ are independent, then the covariance term is null, $E\boldsymbol{\varepsilon}_{dr}(\hat{\boldsymbol{\tau}}_{dr} - \boldsymbol{\tau}_{dr})^T = 0$, and the mean square error of $\hat{\mathbf{y}}_{N_d}$ reduces to $MSE\left(\hat{\mathbf{y}}_{N_d}\right) = MSE(\hat{\boldsymbol{\tau}}_{dr}) + E\boldsymbol{\varepsilon}_{dr}\boldsymbol{\varepsilon}_{dr}^T$, where $E\boldsymbol{\varepsilon}_{dr}\boldsymbol{\varepsilon}_{dr}^T \simeq E\left(\boldsymbol{\varepsilon}_{dr}\boldsymbol{\varepsilon}_{dr}^T\middle|\mathbf{V}_d\right) = \sum_{i_r=1}^{N_d-n_d}\mathbf{V}\mathbf{y}_{di_r}$, where $\mathbf{V}\mathbf{y}_{di_r}$ is the $\mathbf{y}_{di_r}$ covariance matrix.

Appendix 2 for more details on parameter estimations, prediction of random effects, and mean squared error estimations.

4.2.1 Example

We use the field data described in Section 3.2.2. Here, we encode the CTM data as follow: $\mathbf{x}_i = [1 \ 0 \ 0 \ ]$ for a pixel i in the crop-type class maize and other crops, $\mathbf{x}_i = [0 \ 1 \ 0 \ ]$ for crop-type class millet, and $\mathbf{x}_i = [0 \ 0 \ 1 \ ]$ for crop-type class groundnut. We assume that random effects have the same variance σ^2 and are independent among crops: $Cov(v_{dk}, v_{dk'}) = 0$ for $k \neq k'$ and $k = 1, 2, \cdots, K - 1$. The model parameter estimates are in **Table 10**. The values of the z-Wald statistics, for $\boldsymbol{\beta}_k$ parameters, and λ likelihood ratio, for the variance of the random effects σ^2, are shown in parentheses.

According to the z-Wald statistics, the parameters associated with the crop-type classes millet and groundnut are significant for both millet and groundnut (at level < 0.1). The parameters associated with the crop-type class maize and other are not significant (at the 10% level) for both millet and groundnut. According to the likelihood ratio test, the random small-area effect is also significant (at level < 0.01).

Crop-acreage estimates for millet and groundnut in four communities in Senegal's Nioro region are in **Table 11**.

The integration of ground and CTM data makes possible to evaluate the estimates accuracy using the relative root mean squared error, which is the ratio between root mean squared error and crop acreage. It is remarkable that in small areas where the sampling fraction is almost null, this integration reduces the estimation error to levels close to the limits that allow the estimates to be labeled as official: for millet, the estimation error is close to 20%, and for groundnut, the root mean squared error is similar but the acreage is about half the acreage of millet so that, as a result, the relative root mean squared error is about twice than for millet.

5. Optimizing the sample design

The objective is to minimize the sampling variance, that is, maximizing the estimates accuracy, without increasing the cost. The sampling variance depends on the spatial

Crop (k)	$\hat{\boldsymbol{\beta}}_k = \underset{1 \le l \le L}{col}\left(\hat{\beta}_{kl}\right)$ for crop-type class (l)			Variance of the small area effect ($\hat{\sigma}^2$)
	Maize and other	Millet	Groundnut	
Millet	0.109 (z = 0.240)	1.429 (z = 6.236)	0.559 (z = 1.608)	0.028 (λ = 24.636)
Groundnut	−0.662 (z = −1.153)	−0.587 (z = −1.711)	2.522 (z = 8.494)	

Table 10.
Model parameter estimates.

Municipality	Population (pixels)	Sample (pixels)	Crop acreage (hectares) (Error: RRMSE%)	
			Millet	Groundnut
Paos Koto	1,163,948	6	4500 (26.4)	1957 (56.5)
Taiba Niassene	1,163,096	14	4902 (34.3)	2000 (74.5)
Keur Madongo	302,088	2	1281 (24.9)	577 (45.1)
Wack Ngouna	2,766,043	12	9280 (25.0)	4167 (47.1)

Table 11.
Crop acreage estimation at municipality level, Nioro (Senegal).

image to show how the proposed approach should be applied when a CTM is used, instead of a land-use map. The map is digitized, but the minimum level of disaggregation is a square area of side 1 km so that the minimum sampling unit size in the CTM would be a square block of side 100 pixels of 10 m and the survey variable, y_i, would be the number of pixels of the study crop type within sampling unit i.

The studied region was an area of 200 km by 200 km in Castilla y León, Spain, that is, 20,000 by 20,000 pixels of 10 m and $N = 40000$ sampling units of 10,000 pixels. We computed the empirical variogram for nonirrigated herbaceous crops using the moment estimator [17], $2\hat{\gamma}(d) = \frac{1}{|N(d)|}\sum_{N(d)}\left(y_i - y_{i'}\right)^2$, where $N(d) = \{(s_i, s_{i'})$ $|dist((s_i, s_{i'}) = d)\}$. A spherical model was fitted to the empirical variogram and found as parameter estimates, $\hat{\tau}_0 = 728.96$, $\hat{\tau}_d = 539.60$, and $\hat{a} = 46.17$. Using these parameter estimates, we compute the estimate $\Phi(N|\hat{a},\hat{\tau}) = 0.032$. The parameter σ^2 is estimated by $\hat{\sigma}^2 = \frac{S^2}{[1-\Phi(N|\hat{a},\hat{\tau})]}$, where $S^2 = \frac{1}{1-N}\sum_{i=1}^{N}\left(y_i - \bar{y}\right)^2$ and $\bar{y} = \frac{1}{N}\sum_{i=1}^{N} y_i$.

In addition to the sampling unit of minimum size ($n_0 = 1$), we considered a set of sampling unit sizes ranging from $n_0 = 2$ to $n_0 = 10$ times the sampling unit of minimum size ($n_0 = 1$). To estimate $\Phi(n_0|a,\tau)$, it is assumed that the sampling unit of size n_0 is a square lattice of side $\sqrt{n_0}$ and that the distance between a pair of its components is $d = \sqrt{2n_0}$ so that $\Phi(n_0|\hat{a},\hat{\tau}) = (1-\hat{\tau})\left[1 - \frac{3}{2}\frac{\sqrt{2n_0}}{\hat{a}} + \frac{\left(\sqrt{2n_0}\right)^3}{2\hat{a}^3}\right]$.

The optimization problem is

$$\min_{\{n, n_0\}} N^2\left(1-\frac{n}{N}\right)\frac{\hat{\sigma}^2}{n}\frac{n_0(Nn_0-1)}{(N-1)}[1-0.032] - \frac{Nn_0(n_0-1)}{(N-1)}\left[1-(1-\hat{\tau})\left(1-\frac{3}{2}\frac{\sqrt{2n_0}}{\hat{a}}+\frac{\left(\sqrt{2n_0}\right)^3}{2\hat{a}^3}\right)\right].$$

$$s.t. \quad C_0 + C_w n n_0 + C_k\sqrt{An} \leq C$$

To find the solution to this problem, we use the cost function coefficients in [21] ($C = 120000\$$, $C_0 = 20500\$$, $C_w = 50\$$, and $C_k = 1\$/Km$) and the nloptr package [22]. The solution is in **Table 12**.

The optimum sampling unit size is the lower bound allowed. The minimum anticipated variance increases from 1.1 e+9 to 3.8e+9, when the sampling unit size increases from 1 to 10. In other words, using sampling units of size 10, the sampling variance is 1.86 times higher than using sampling units of size 1.

6. Concluding remarks

We have shown how to integrate CTM and ground data to produce agricultural and environmental statistics, along with measures of their accuracy that take into

Variable	Optimal solution					
	Lower bound for the sampling unit size					
	1	2	3	4	5	10
Sample size: n	1661	891	611	465	375	192
Sampling unit size: n_0	1	2	3	4	5	10
Minimum Anticipated Variance	1.1 e+9	1.4 e+9	1.7e+9	2.0e+9	2.3e+9	3.8e+9

Table 12.
SRS: Optimum solution for several values of lower bound for the sampling unit size.

account any source of uncertainty, including classification errors. We have developed prototypes for this integration at the national, provincial, and municipal levels. For large areas (national and provincial level), these prototypes are based on the design of a probabilistic scheme to select the sample from which to collect the ground data, and we use a statistical model to integrate this sample with CTM data. The resulting statistics are design-consistent and robust; that is, they are based on the sampling design rather than on the model. The sample design is key, and we have shown how to optimize it using CTM. For small areas (municipal level), the prototypes are based solely on the model so that the resulting statistics are model-dependent.

To facilitate the transference of these prototypes, we have included examples of their application to data provided by two National Statistical Offices: Spain and Senegal. The former is an example of countries where ground data are collected at the parcel level, and we use linear models for their integration with CTM. The latter is an example of countries where ground data are observed at the pixel level, and we use multinomial logit models for their integration with CTM. We show how to estimate the model parameters and how to use CTM data to compute the required statistics, as well as their accuracy measures in these two countries, at both large and small area levels.

We have evaluated the efficiency of CTM data to improve the methods currently used to produce official agricultural statistics in these two countries, and we have found it to be high: using CTM data, the current sample size at the level of large areas could be reduced to less than one-fifth in the case of Spain and between one-third and one-fifth in the case of Senegal, without loss of neither robustness nor accuracy. At the small area level, where the sampling fraction is almost null, we have shown how to produce estimates accurate enough to be labeled as official statistics, that is, useful for most applications, without increasing costs.

Acknowledgements

The research results presented in this chapter were supported by the European Space Agency, in the Sen4Stat project framework [Contract No. 4000127181/19/I-NS]. The national statistical offices of Senegal and Spain provided the ground data used in the examples.

A. Appendix 1—pixels classification

We want to divide the $i = 1, 2, \cdots, N$ pixels into a set of $k = 1, 2, \cdots, K$ groups, each one corresponding to a discriminable class on the terrain identified by a crop type or land use. Let $f(i \in k, \mathbf{x}_i)$ be the joint probability function of the events "pixel i is in the terrain class k $(i \in k)$, and the spectral measure in i is $\mathbf{x}_i$". The decision rule is to "classify i in the class k if and only if $f(i \in k, \mathbf{x}_i) = \max\limits_{k=1,2,\cdots,K} f(i \in k, \mathbf{x}_i)$." Maximum likelihood, maximum entropy, and multinomial regression are three approaches to implement this rule.

A.1 Maximum likelihood

This approach looks for minimizing the expected loss due to classification errors. Let $\lambda_i(k'|k)$ be a loss function measuring the cost incurred when pixel i is classified in class $k' \neq k$ but is actually of class k. The expected loss is $E\lambda_i(k'|k) = \sum_{k'=1}^{K}\lambda_i(k'|k)f(i\in k'|\mathbf{x}_i)$, where $f(i\in k'|\mathbf{x}_i) = f(i\in k')f(\mathbf{x}_i|i\in k')/f(\mathbf{x}_i)$. Using this loss function, we consider the set of K discriminant functions, $\left\{g_{k'}(\mathbf{x}_i) = -\sum_{k'=1}^{K}\lambda_i(k'|k)f(i\in k')f(\mathbf{x}_i|i\in k')/f(\mathbf{x}_i); i = 1, 2, \cdots, N; k' = 1, 2, \cdots, K\right\}$ and we decide to classify the pixel i in the class k' if $g_{k'}(\mathbf{x}_i) = \max_{k=1,2,\cdots,K} g_{k'}(\mathbf{x}_i) = \min_{k=1,2,\cdots,K} -g_{k'}(\mathbf{x}_i)$. Note that this classifier minimizes the expected loss, as desired. A simple loss function is $\lambda_i(k'|k) = 0 \text{ if } k = k'$ and $\lambda_i(k'|k) = 1$ otherwise. Using this loss function, the discriminant function is $g_{k'}(\mathbf{x}_i) = -\sum_{k'(\neq k)=1}^{K} f(\mathbf{x}_i|i\in k')f(i\in k')/f(\mathbf{x}_i)$, and since $f(\mathbf{x}_i)$ is a constant for a given $\mathbf{x}_i$, an equivalent discriminant function is $g'_{k'}(\mathbf{x}_i) = g_{k'}(\mathbf{x}_i)f(\mathbf{x}_i) = -\sum_{k'(\neq k)=1}^{K} f(\mathbf{x}_i|i\in k')f(i\in k')$. Note that $f(\mathbf{x}_i) = \sum_{k'=1}^{K} f(i\in k')f(\mathbf{x}_i|i\in k') = \sum_{k'(\neq k)=1}^{K} f(\mathbf{x}_i|i\in k')f(i\in k') + f(\mathbf{x}_i|i\in k)f(i\in k)$, and as a result, $g'_{k'}(\mathbf{x}_i) = -f(\mathbf{x}_i) + f(\mathbf{x}_i|i\in k)f(i\in k)$. Since $f(\mathbf{x}_i)$ is a constant for a given $\mathbf{x}_i$, an equivalent discriminant function is $g''_{k'}(\mathbf{x}_i) = g'_{k'}(\mathbf{x}_i) + f(\mathbf{x}_i) = f(\mathbf{x}_i|i\in k)f(i\in k)$.

Thus, we consider the set of K discriminant functions, $\left\{g''_{k'}(\mathbf{x}_i) = f(\mathbf{x}_i|i\in k)f(i\in k); i = 1, 2, \cdots, N; k = 1, 2, \cdots, K\right\}$ and we decide to classify the pixel i in the class k' if $g''_{k'}(\mathbf{x}_i) = \max_{k=1,2,\cdots,K} g''_{k'}(\mathbf{x}_i) = \max_{k=1,2,\cdots,K} f(\mathbf{x}_i|i\in k)f(i\in k)$.

Note that $\max_{k=1,2,\cdots,K} f(\mathbf{x}_i|i\in k)f(i\in k) = \max_{k=1,2,\cdots,K} \log f(\mathbf{x}_i|i\in k)f(i\in k)$, where $\log f(\mathbf{x}_i|i\in k)f(i\in k) = \log f(i\in k) + \log f(\mathbf{x}_i|i\in k)$, where $\log f(\mathbf{x}_i|i\in k) = -\frac{L}{2}\log(2\pi) - \frac{1}{2}\log|\det\mathbf{V}_k| - \left[\frac{1}{2}(\mathbf{x}_i - \boldsymbol{\mu}_k)\mathbf{V}_k^{-1}(\mathbf{x}_i - \boldsymbol{\mu}_k)^T\right]$. Thus, an equivalent set of K discriminant functions is $\left\{g'''_k(\mathbf{x}_i) = \log f(i\in k) - \frac{1}{2}\log|\det\mathbf{V}_k| - \left[\frac{1}{2}(\mathbf{x}_i - \boldsymbol{\mu}_k)\mathbf{V}_k^{-1}(\mathbf{x}_i - \boldsymbol{\mu}_k)^T\right]; i = 1, 2, \cdots N, k = 1, 2, \cdots K\right\}$ and we decide to classify the pixel i in the class k if $g'''_k(\mathbf{x}_i) = \max_{k=1,2,\cdots,K} g'''_k(\mathbf{x}_i)$.

A.2 Maximum entropy

a. **Multinomial distribution**. The entropy $H(\boldsymbol{\mu})$ can be derived from the multinomial distribution. Assume that the survey vector, $\mathbf{y}_i = \underset{1\le k\le K}{col}(y_{ik})$, follows a multinomial distribution, $MN(1; \boldsymbol{\mu}_i)$. Then, $\sum_{i=1}^{n}\mathbf{y}_i = \underset{1\le k\le K}{col}\left(y_k = \sum_{i=1}^{n} y_{ik}\right)$ follows a multinomial distribution $MN\left(n = \sum_{k=1}^{K} y_k; \boldsymbol{\mu} = \underset{1\le k\le K}{col}(\mu_k)\right)$, with $\sum_{k=1}^{K}\mu_k = 1$. The number of $\sum_{i=1}^{n}\mathbf{y}_i$ realizations favorable to the observed sample $\mathbf{y}_n = \underset{1\le k\le K}{col}(y_k)$ is $W = \frac{n!}{\prod_{k=1}^{K} y_k!}$. Thus, $\ln W = \ln n! - \sum_{k=1}^{K}\ln y_k!$ and, using the Stirling approximation, $\ln W = n\ln n - n - \sum_{k=1}^{K} y_k \ln y_k + n = n\ln n - \sum_{k=1}^{K} y_k \ln y_k$. For large samples, and according to the Bernoulli theorem, $\frac{y_k}{n} \to \mu_k$ when $n \to \infty$, and as a result,

DOI: http://dx.doi.org/10.5772/intechopen.1004409

$\ln W \approx n \ln n - \sum_{k=1}^{K} n\mu_k \ln n\mu_k = n \ln n - n\left(\sum_{k=1}^{K} \mu_k (\ln n + \ln \mu_k)\right) =$
$-n\sum_{k=1}^{K} \mu_k \ln \mu_k$. Finally, $n^{-1} \ln W \approx -\sum_{k=1}^{K} \mu_k \ln \mu_k = H(\boldsymbol{\mu})$. The entropy and multinomial logit approaches provide coincident class probability estimates because they are maximizing the same function.

b. **Information gain measures**. In the simple case where the available information reduces to sample ground data, $\mathbf{y} = \underset{1 \le i \le n}{col} \underset{1 \le k \le K}{col} (y_{ik})$, $\boldsymbol{\mu}$ is considered compatible with the observed data $\mathbf{y}_n = \sum_{i=1}^{n} \mathbf{y}_i = \underset{1 \le k \le K}{col} \left(y_k = \sum_{i=1}^{n} y_{ik}\right)$, when $\mathbf{y}_n = \boldsymbol{\mu} + \boldsymbol{\varepsilon}_n$, where $\boldsymbol{\varepsilon}_n = \underset{1 \le k \le K}{col} \left(\sum_{i=1}^{n} \varepsilon_{ik}\right)$ and ε_{ik} is a random noise. The set of $\boldsymbol{\mu} = \underset{1 \le k \le K}{col} (\mu_k)$ values compatible with $\mathbf{y}_n$ is countless, and according to the *maximum entropy* principle, we choose as $\boldsymbol{\mu}$ value the one making $H(\boldsymbol{\mu})$ maximum: $\hat{\boldsymbol{\mu}} = \max_{\boldsymbol{\mu}} H(\boldsymbol{\mu}) = \max_{\boldsymbol{\mu}} -\sum_{k=1}^{K} \mu_k \ln \mu_k = \min_{\boldsymbol{\mu}} \sum_{k=1}^{K} \mu_k \ln \mu_k$, subject to $\sum_{k=1}^{K} \mu_k = 1$. To find $\hat{\boldsymbol{\mu}}$, we solve the Lagrange function $\Psi = -\sum_{k=1}^{K} \mu_k \ln \mu_k + \lambda\left(1 - \sum_{k=1}^{K} \mu_k\right)$. The Ψ derivative with respect to μ_k is $\frac{\partial \Psi}{\partial \mu_k} = \frac{\partial}{\partial \mu_k}\left[-\sum_{k=1}^{K} \mu_k \ln \mu_k + \lambda\left(1 - \sum_{k=1}^{K} \mu_k\right)\right] =$
$-\left(\mu_k \frac{1}{\mu_k} + \ln \mu_k\right) - \lambda = -\ln \mu_k - 1 - \lambda$, for $k = 1, 2, \cdots, K$ and with respect to λ is $\frac{\partial \Psi}{\partial \lambda} = \left(1 - \sum_{k=1}^{K} \mu_k\right)$. The solution of this system of $K+1$ equations is $-\ln \hat{\mu}_k - 1 - \hat{\lambda} = 0 \Leftrightarrow \hat{\mu}_k = e^{-1-\hat{\lambda}}; k = 1, 2, \cdots, K$, with $\sum_{k=1}^{K} \hat{\mu}_k = Ke^{-1-\hat{\lambda}} = 1$, so that $\ln Ke^{-1-\hat{\lambda}} = \ln 1 = 0 \Leftrightarrow \ln K - 1 - \hat{\lambda} = 0 \Leftrightarrow \hat{\lambda} = \ln K - 1$. The entropy is $H(\hat{\boldsymbol{\mu}}) = -\sum_{k=1}^{K} e^{-1-\hat{\lambda}}(-1 - \hat{\lambda}) = Ke^{-1-\hat{\lambda}}(1 + \hat{\lambda}) = (1 + \hat{\lambda}) = \ln K$, and it is achieved when $\hat{\mu}_k = e^{-1-\hat{\lambda}} = e^{-\ln K}; k = 1, 2, \cdots, K$, that is, when, $\ln \hat{\mu}_k =$
$-\ln K = \ln \frac{1}{K} \Leftrightarrow \hat{\mu}_k = \frac{1}{K}; k = 1, 2, \cdots, K$ so that the assigned probabilities $\hat{\boldsymbol{\mu}} = \underset{1 \le k \le K}{col} (\hat{\mu}_k)$ correspond to a uniform distribution.

When additional information, $\mathbf{X} = \underset{1 \le i \le n}{col} \underset{1 \le l \le L}{row} (x_{il})$, is introduced, then a reduction of uncertainty is achieved, from $\ln K$ to $-\sum_{k=1}^{K} \hat{\mu}_k \ln \hat{\mu}_k$. This reduction, $\ln K + \sum_{k=1}^{K} \hat{\mu}_k \ln \hat{\mu}_k$, is called "information gain," and the ratio between this reduction and the maximum uncertainty, $I(\hat{\boldsymbol{\mu}}) = \frac{\ln K + \sum_{k=1}^{K} \hat{\mu}_k \ln \hat{\mu}_k}{\ln K} = 1 - S(\hat{\boldsymbol{\mu}})$, is called "information index", where $S(\hat{\boldsymbol{\mu}}) = -\frac{\sum_{k=1}^{K} \hat{\mu}_k \ln \hat{\mu}_k}{\ln K}$. A value $S(\hat{\boldsymbol{\mu}}) = 0$, that is, $I(\hat{\boldsymbol{\mu}}) = 1$, corresponds to no uncertainty ($\hat{\mu}_k = 1$ for some k and $\hat{\mu}_{k'} = 0$ for every $k' \neq k$), and a value $S(\hat{\boldsymbol{\mu}}) = 1$, that is, $I(\hat{\boldsymbol{\mu}}) = 0$, corresponds to perfect uncertainty ($\hat{\mu}_k = \frac{1}{K}; k = 1, 2, \cdots, K$). The measure $S(\hat{\boldsymbol{\mu}})$ can be used to evaluate the effect of each piece of information. For instance, the effect of reducing the sample size in a unit can be measured by comparing $S(\hat{\boldsymbol{\mu}}|n)$ and $S(\hat{\boldsymbol{\mu}}|n-1)$: if $S(\hat{\boldsymbol{\mu}}|n) < S(\hat{\boldsymbol{\mu}}|n-1)$, then the unit in question reduced the uncertainty. In the same way, the effect of excluding an auxiliary variable can be measured using $S(\hat{\boldsymbol{\mu}}|L)$ and $S(\hat{\boldsymbol{\mu}}|L-1)$: if $S(\hat{\boldsymbol{\mu}}|L) < S(\hat{\boldsymbol{\mu}}|L-1)$, then the variable in question reduces the uncertainty.

B. Appendix 2 - Small area estimation based on multinomial logit mixed models

B.1 Parameter estimation

To estimate μ_{dik}, it is sufficient to acquire estimates of $\boldsymbol{\beta}_k$ for $k = 1, 2, \cdots, K-1$. The probability that crop K covers pixel i of d is $\mu_{diK} = 1 - \sum_{k=1}^{K-1} \mu_{dik}$ and can be estimated using the estimates of $\boldsymbol{\beta}_k$ for $k = 1, 2, \cdots, K-1$. We specify the vector $\boldsymbol{\eta} = \underset{1 \le d \le D}{col} \underset{1 \le i \le N_d}{col} \underset{1 \le k \le K-1}{col} (\eta_{dik})$ where $\eta_{dik} = \mathbf{x}_{di}\boldsymbol{\beta}_k + v_{dk}$ as a linear mixed model $\boldsymbol{\eta} = \mathbf{X}\boldsymbol{\beta} + \mathbf{Z}\mathbf{v}$, where $\mathbf{X} = \underset{1 \le d \le D}{col} \underset{1 \le i \le N_d}{col} \underset{1 \le k \le K-1}{diag} \underset{1 \le l \le L}{row} (x_{dil})$ are known CTM values, $\boldsymbol{\beta} = \underset{1 \le k \le K-1}{col} \underset{1 \le l \le L}{col} (\beta_{kl})$ is a vector of parameters, and $\mathbf{Z} = \underset{1 \le d \le D}{diag} \underset{1 \le i \le N_d}{col} \underset{1 \le k \le K-1}{row} (z_{dik})$ is an indicator matrix where $z_{dik} = 1$ if k covers i of d and $z_{dik} = 0$ otherwise. It is assumed that the vector of small-area random effect $\mathbf{v} = \underset{1 \le d \le D}{col} \underset{1 \le k \le K-1}{col} (v_{dk})$ has mean $E\mathbf{v} = \mathbf{0}$ and covariance matrix $\boldsymbol{\Sigma} = \boldsymbol{\theta} \otimes \mathbf{I}_D$, where $\boldsymbol{\theta} = \underset{1 \le k \le K-1}{row} \underset{1 \le k' \le K-1}{col} \left(\sigma^2_{kk'}\right)$ and $\sigma^2_{kk'} = \sigma^2_k$ if $k = k'$.

The parameters in $\boldsymbol{\beta}$ and $\boldsymbol{\theta}$ are often estimated following a penalized quasi-likelihood approach [22–24]. As discussed by McCulloch and Searle [13], these methods are not completely satisfactory in practice, and these authors recommended to linearize the nonlinear mixed model instead, thus obtaining a working linear mixed model fitted by maximum likelihood. Here, we follow this recommendation and consider the link function applied to the ground data $g_k(\mathbf{y}_{di}) = \log\left(\frac{y_{dik}}{1 - \sum_{k=1}^{K-1} y_{dik}}\right)$ for $k = 1, 2, \cdots, K-1$, and its Taylor expansion about $E(\mathbf{y}_{di}|\mathbf{v}) = \boldsymbol{\mu}_{di}$, $g_k(\mathbf{y}_{di}) \simeq g_k(\boldsymbol{\mu}_{di}) + \left.\frac{\partial g_k(\mathbf{y}_{di})}{\partial y_{di1}}\right|_{\mathbf{y}_{di} = \boldsymbol{\mu}_{di}} (y_{di1} - \mu_{di1}) + \cdots + \left.\frac{\partial g_k(\mathbf{y}_{di})}{\partial y_{diK-1}}\right|_{\mathbf{y}_{di} = \boldsymbol{\mu}_{di}} (y_{diK-1} - \mu_{diK-1})$, to obtain the working linear model $\boldsymbol{\xi} = \mathbf{X}\boldsymbol{\beta} + \mathbf{Z}\mathbf{v} + \mathbf{e}$ [24]. Here, $\boldsymbol{\xi} = \underset{1 \le d \le D}{col} \underset{1 \le i \le N_d}{col} (\boldsymbol{\xi}_{di})$ and $\boldsymbol{\xi}_{di} = \underset{1 \le k \le K-1}{col} (\xi_{dik})$ is a working vector, $\mathbf{e} = \underset{1 \le d \le D}{col} \underset{1 \le i \le N_d}{col} (\mathbf{e}_{di})$ with $\mathbf{e}_{di} = \mathbf{H}_{di}^{-1}(\mathbf{y}_{di} - \boldsymbol{\mu}_{di})$, and $\mathbf{H}_{di} = \frac{\partial h(\boldsymbol{\eta}_{di})}{\partial \boldsymbol{\eta}_{di}} = \underset{1 \le k \le K-1}{col} \underset{1 \le k' \le K-1}{row} (\delta_{kk'}\mu_{dik} - \mu_{dik}\mu_{dik'})$ with $\delta_{kk'} = 1$ if $k = k'$ and $\delta_{kk'} = 0$ otherwise, $\mathbf{y}_{di} = \underset{1 \le k \le K-1}{col} (y_{dik})$, and $\boldsymbol{\mu}_{di} = \underset{1 \le k \le K-1}{col} (\mu_{dik})$. Here, $h(\boldsymbol{\eta}_{di})$ is the inverse of the link function $g(\boldsymbol{\mu}_{di}) = \boldsymbol{\eta}_{di}$, that is, $\boldsymbol{\mu}_{di} = g^{-1}(\boldsymbol{\eta}_{di}) = h(\boldsymbol{\eta}_{di})$.

The working vector covariance matrix is $\mathbf{V} = \mathbf{V}\boldsymbol{\xi} = \mathbf{Z}\boldsymbol{\Sigma}\mathbf{Z}^T + \mathbf{V}\mathbf{e}$, where $\mathbf{V}\mathbf{e} = \underset{1 \le d \le D}{diag} \underset{1 \le i \le N_d}{diag} (\mathbf{V}\mathbf{e}_{di})$ and $\mathbf{V}\mathbf{e}_{di} = \mathbf{H}_{di}^{-1}\mathbf{V}(\mathbf{y}_{di} - \boldsymbol{\mu}_{di})\mathbf{H}_{di}^{-1}$, where $\mathbf{V}(\mathbf{y}_{di} - \boldsymbol{\mu}_{di})$ is the covariance matrix of $\mathbf{y}_{di} \to MN(1, \boldsymbol{\mu}_{di})$.

We assume that the $\boldsymbol{\xi}$ asymptotic distribution is normal, $\boldsymbol{\xi} \to AN(\mathbf{X}\boldsymbol{\beta}, \mathbf{V})$, and we see $\boldsymbol{\xi} = \mathbf{X}\boldsymbol{\beta} + \mathbf{Z}\mathbf{v} + \mathbf{e}$ as a linear mixed model. $\mathbf{V}\mathbf{e}_{di}$ is complex because the derivative of the link function $\mathbf{H}_{di}$ depends on $\mathbf{V}$ through $\boldsymbol{\mu}$, and $\boldsymbol{\mu}_{di}$ also depends on $\mathbf{V}$. We consider a simplification of this problem; it consists of setting $\mathbf{v} = \mathbf{0}$ in $\boldsymbol{\mu}_{di}$ that reduces $\mathbf{H}_{di}$ to $\mathbf{H}^*_{di} = \underset{1 \le k \le K-1}{col} \underset{1 \le k' \le K-1}{row} \left(\delta_{kk'}\mu^*_{dik} - \mu^*_{dik}\mu^*_{dik'}\right)$, where $\mu^*_{dik} = \frac{\exp(\mathbf{x}_{di}\boldsymbol{\beta}_k)}{1 + \sum_{k=1}^{K-1} \exp(\mathbf{x}_{di}\boldsymbol{\beta}_k)}$ no longer depends on the random effects $\mathbf{v}$ [13]. As a result, $\mathbf{Ve}^*_{di} = \mathbf{H}^{*-1}_{di}\mathbf{V}(\mathbf{y}_{di} - \boldsymbol{\mu}_{di})\mathbf{H}^{*-1}_{di}$, where $\boldsymbol{\mu}_{di}$ does depend on $\mathbf{v}$.

To determine the $\boldsymbol{\beta}$ and $\boldsymbol{\theta}$ maximum likelihood estimates, we follow an iterative procedure:

1. Let $\boldsymbol{\theta}^{(0)}$ and $\boldsymbol{\beta}^{(0)}$ be starting values of $\boldsymbol{\theta}$ and $\boldsymbol{\beta}$, respectively

2. Let $\mathbf{v}^{(0)} = 0$ and compute $\boldsymbol{\xi}^{(0)} = \mathbf{X}\boldsymbol{\beta}^{(0)} + \mathbf{H}^{*-1(0)}\left(\mathbf{y} - \boldsymbol{\mu}^{(0)}\right)$

3. Compute $\mathbf{V}\boldsymbol{\xi}^{(0)} = \underset{1\le d\le D}{diag}\left(\mathbf{Ve}^{*(0)}_{di}\right)$, where $\mathbf{Ve}^{*(0)}_{di} = \mathbf{H}^{*-1(0)}_{di}\mathbf{V}\left(\mathbf{y}_{di} - \boldsymbol{\mu}^{(0)}_{di}\right)\mathbf{H}^{*-1(0)}_{di}$ with $\boldsymbol{\mu}^{(0)}_{di} = \underset{1\le k\le K-1}{col}\left(\mu^{(0)}_{dik}\right)$ and $\mu^{(0)}_{dik} = \frac{\exp\left(\mathbf{x}_{di}\boldsymbol{\beta}^{(0)}_k\right)}{1+\sum_{k=1}^{K-1}\exp\left(\mathbf{x}_{di}\boldsymbol{\beta}^{(0)}_k\right)}$

4. Update $\boldsymbol{\beta}^{(0)}$ using $\hat{\boldsymbol{\beta}}^{(1)} = \left(\mathbf{X}^T\mathbf{V}\boldsymbol{\xi}^{(0)^{-1}}\mathbf{X}\right)^{-1}\mathbf{X}^T\mathbf{V}\boldsymbol{\xi}^{(0)^{-1}}\boldsymbol{\xi}^{(0)}$

5. Update $\boldsymbol{\theta}^{(0)}$ using $\hat{\boldsymbol{\theta}}^{(1)} = \boldsymbol{\theta}^{(0)} - \left[\frac{\partial^2 l(\boldsymbol{\beta})}{\partial\boldsymbol{\theta}\partial\boldsymbol{\theta}^T}\right]^{-1}_{\boldsymbol{\beta}=\hat{\boldsymbol{\beta}}^{(1)}}\frac{\partial l(\boldsymbol{\beta})}{\partial\boldsymbol{\theta}}\bigg|_{\boldsymbol{\beta}=\hat{\boldsymbol{\beta}}^{(1)}}$, where $l(\boldsymbol{\beta})$ is the profile log likelihood of the $\boldsymbol{\xi}$ distribution.

6. Update $\hat{\mathbf{V}}^{(0)}$ using $\hat{\mathbf{v}}^{(1)} = \hat{\boldsymbol{\Sigma}}^{(1)}\mathbf{Z}^T\mathbf{V}\boldsymbol{\xi}^{(1)^{-1}}\left(\boldsymbol{\xi}^{(0)} - \mathbf{X}\hat{\boldsymbol{\beta}}^{(1)}\right)$ and compute $\boldsymbol{\xi}^{(1)} = \mathbf{X}\hat{\boldsymbol{\beta}}^{(1)} + \mathbf{Z}\hat{\mathbf{v}}^{(1)} + \hat{\mathbf{H}}^{*-1(1)}\left(\mathbf{y} - \hat{\boldsymbol{\mu}}^{(1)}\right)$

7. Update $\hat{\boldsymbol{\beta}}^{(m)}$ using $\hat{\boldsymbol{\beta}}^{(m+1)} = \left(\mathbf{X}^T\mathbf{V}\boldsymbol{\xi}^{(m)^{-1}}\mathbf{X}\right)^{-1}\mathbf{X}^T\mathbf{V}\boldsymbol{\xi}^{(m)^{-1}}\boldsymbol{\xi}^{(m)}$

8. Update $\hat{\boldsymbol{\theta}}^{(m)}$ using $\hat{\boldsymbol{\theta}}^{(m+1)} = \hat{\boldsymbol{\theta}}^{(m)} - \left[\frac{\partial^2 l(\boldsymbol{\beta})}{\partial\boldsymbol{\theta}\partial\boldsymbol{\theta}^T}\right]^{-1}_{\boldsymbol{\beta}=\hat{\boldsymbol{\beta}}^{(m)}}\frac{\partial l(\boldsymbol{\beta})}{\partial\boldsymbol{\theta}}\bigg|_{\boldsymbol{\beta}=\hat{\boldsymbol{\beta}}^{(m)}}$

The gradient vector for $\boldsymbol{\theta} = \underset{1\le k\le K-1}{row}\ \underset{1\le k'\le K-1}{col}\left(\sigma^2_{kk'}\right)$ is $\mathbf{b} = \frac{\partial l(\boldsymbol{\beta})}{\partial\boldsymbol{\theta}} = \underset{1\le l\le L}{col}\left(\frac{\partial l}{\partial\theta_l}\right)$, where $\theta_l = \sigma_k$ for $k = k'$ and $\theta_l = \sigma^2_{kk'}$ for $k \ne k'$ and $\frac{\partial l}{\partial\theta_l} = -\frac{1}{2}tr\left(\mathbf{V}^{-1}\mathbf{V}_l\right) + \frac{1}{2}\mathbf{r}^T\left(\mathbf{V}^{-1}\mathbf{V}_l\mathbf{V}^{-1}\right)\mathbf{r}$, where $\mathbf{r}^{(m+1)} = \boldsymbol{\xi}^{(m)} - \mathbf{X}\boldsymbol{\beta}^{(m+1)}$ and $\mathbf{V}_l = \frac{\partial\mathbf{V}}{\partial\theta_l}$, which for $\theta_l = \sigma_k$ is equal to $\mathbf{V}_{\sigma_k} = \frac{\partial\mathbf{V}}{\partial\sigma_k} = \mathbf{Z}\left(\mathbf{I}_D \otimes \frac{\partial\boldsymbol{\theta}}{\partial\sigma_k}\right)\mathbf{Z}^T$ with $\frac{\partial\boldsymbol{\theta}}{\partial\sigma_k} = \underset{1\le k\le K-1}{diag}(2\sigma_k)$ and for $\theta_l = \sigma^2_{kk'}$ when $k \ne k'$ it is $\mathbf{V}_{\sigma^2_{kk'}} = \frac{\partial\mathbf{V}}{\partial\sigma^2_{kk'}} = \mathbf{Z}\left(\mathbf{I}_D \otimes \frac{\partial\boldsymbol{\theta}}{\partial\sigma^2_{kk'}}\right)\mathbf{Z}^T$ with $\frac{\partial\boldsymbol{\theta}}{\partial\sigma^2_{kk'}} = \frac{\partial\ \underset{1\le k\le K-1}{row}\ \underset{1\le k'\le K-1}{col}\left(\sigma^2_{kk'}\right)}{\partial\sigma^2_{kk'}}$.

The Hessian matrix for $\boldsymbol{\theta}$ is $\frac{\partial^2 l(\boldsymbol{\beta})}{\partial\boldsymbol{\theta}\partial\boldsymbol{\theta}^T} = \underset{1\le l\le L}{row}\ \underset{1\le l'\le L}{col}\left(\frac{\partial^2 l}{\partial\theta_l\partial\theta_{l'}}\right)$, where we use l instead of $l(\boldsymbol{\beta})$. Here, $\frac{\partial^2 l}{\partial\theta_l\partial\theta_{l'}} = -\frac{1}{2}tr\left(\mathbf{V}^{-1}\mathbf{V}_{ll'} - \mathbf{V}^{-1}\mathbf{V}_l\mathbf{V}^{-1}\mathbf{V}_{l'}\right) - \frac{1}{2}\mathbf{e}^T\mathbf{V}^{ll'}\mathbf{e}$ with $\mathbf{V}_{ll'} = \frac{\partial^2\mathbf{V}}{\partial\theta_l\partial\theta_{l'}}$ and $\mathbf{V}^{ll'} = \frac{\partial^2\mathbf{V}^{-1}}{\partial\theta_l\partial\theta_{l'}} = \mathbf{V}^{-1}\left(\mathbf{V}_{ll'} - \mathbf{V}_l\mathbf{V}^{-1}\mathbf{V}_{l'} - \mathbf{V}_{l'}\mathbf{V}^{-1}\mathbf{V}_l\right)\mathbf{V}^{-1}$.

For $K = 3$ and $l, l' = \sigma_1, \sigma_2, \sigma_{12}^2$, the Hessian matrix is

$$\frac{\partial^2 l(\boldsymbol{\beta})}{\partial \boldsymbol{\theta} \partial \boldsymbol{\theta}^T} = \begin{bmatrix} \frac{\partial^2 l}{\partial \sigma_1 \partial \sigma_1} & \frac{\partial^2 l}{\partial \sigma_1 \partial \sigma_2} & \frac{\partial^2 l}{\partial \sigma_1 \partial \sigma_{12}^2} \\ \frac{\partial^2 l}{\partial \sigma_2 \partial \sigma_1} & \frac{\partial^2 l}{\partial \sigma_2 \partial \sigma_2} & \frac{\partial^2 l}{\partial \sigma_2 \partial \sigma_{12}^2} \\ \frac{\partial^2 l}{\partial \sigma_{12}^2 \partial \sigma_1} & \frac{\partial^2 l}{\partial \sigma_{12}^2 \partial \sigma_2} & \frac{\partial^2 l}{\partial \sigma_{12}^2 \partial \sigma_{12}^2} \end{bmatrix}, \text{ and}$$

$$\mathbf{V}_{\sigma_1\sigma_1} = \frac{\partial^2 \mathbf{V}}{\partial \sigma_1 \partial \sigma_1} = \mathbf{Z}\left(\mathbf{I}_D \otimes \frac{\partial^2 \boldsymbol{\theta}}{\partial \sigma_1 \partial \sigma_1}\right)\mathbf{Z}^T = \mathbf{Z}\left(\mathbf{I}_D \otimes \begin{bmatrix} 2 & 0 \\ 0 & 0 \end{bmatrix}\right)\mathbf{Z}^T,$$

$$\mathbf{V}_{\sigma_1\sigma_2} = \frac{\partial^2 \mathbf{V}}{\partial \sigma_1 \partial \sigma_2} = \frac{\partial^2 \mathbf{V}}{\partial \sigma_1 \partial \sigma_2} = \mathbf{Z}\left(\mathbf{I}_D \otimes \frac{\partial^2 \boldsymbol{\theta}}{\partial \sigma_1 \partial \sigma_2}\right)\mathbf{Z}^T = \mathbf{Z}\left(\mathbf{I}_D \otimes \begin{bmatrix} 0 & 0 \\ 0 & 0 \end{bmatrix}\right)\mathbf{Z}^T = \mathbf{0}_{n(K-1)\times n(K-1)},$$

$$\mathbf{V}_{\sigma_1\sigma_{12}^2} = \frac{\partial^2 \mathbf{V}}{\partial \sigma_1 \partial \sigma_{12}^2} = \mathbf{Z}\left(\mathbf{I}_D \otimes \frac{\partial^2 \boldsymbol{\theta}}{\partial \sigma_1 \partial \sigma_{12}^2}\right)\mathbf{Z}^T = 0_{n(K-1)\times n(K-1)}, \mathbf{V}_{\sigma_2\sigma_1} = \mathbf{V}_{\sigma_1\sigma_2}$$

$$\mathbf{V}_{\sigma_2\sigma_2} = \frac{\partial^2 \mathbf{V}}{\partial \sigma_2 \partial \sigma_2} = \mathbf{Z}\left(\mathbf{I}_D \otimes \frac{\partial^2 \boldsymbol{\theta}}{\partial \sigma_2 \partial \sigma_2}\right)\mathbf{Z}^T = \mathbf{Z}\left(\mathbf{I}_D \otimes \begin{bmatrix} 0 & 0 \\ 0 & 2 \end{bmatrix}\right)\mathbf{Z}^T,$$

$$\mathbf{V}_{\sigma_2\sigma_{12}^2} = \frac{\partial^2 \mathbf{V}}{\partial \sigma_2 \partial \sigma_{12}^2} = \mathbf{Z}\left(\mathbf{I}_D \otimes \frac{\partial^2 \boldsymbol{\theta}}{\partial \sigma_2 \partial \sigma_{12}^2}\right)\mathbf{Z}^T = \mathbf{0}_{n(K-1)\times n(K-1)}, \mathbf{V}_{\sigma_{12}^2\sigma_1} = \mathbf{V}_{\sigma_1\sigma_{12}^2},$$

$$\mathbf{V}_{\sigma_{12}^2\sigma_2} = \mathbf{V}_{\sigma_2\sigma_{12}^2}, \mathbf{V}_{\sigma_{12}^2\sigma_{12}^2} = \frac{\partial^2 \mathbf{V}}{\partial \sigma_{12}^2 \partial \sigma_{12}^2} = \mathbf{Z}\left(\mathbf{I}_D \otimes \frac{\partial^2 \boldsymbol{\theta}}{\partial \sigma_{12}^2 \partial \sigma_{12}^2}\right)\mathbf{Z}^T = \mathbf{0}_{n(K-1)\times n(K-1)}$$

9. Continue until convergence.

B.2 Prediction

We partition the linear predictor vector $\boldsymbol{\eta}_d = \underset{1\le i\le N_d}{col}\ \underset{1\le k\le K-1}{col}(\eta_{dik})$ into a part $\boldsymbol{\eta}_{ds} = \underset{1\le i_s\le n_d}{col}\ \underset{1\le k\le K-1}{col}(\eta_{di_sk})$ containing the n_d known sample values and another part $\boldsymbol{\eta}_{dr} = \underset{1\le i_r\le N_d-n_d}{col}\ \underset{1\le k\le K-1}{col}(\eta_{di_rk})$ containing $N_d - n_d$ unknown values we want to predict. We partition the model $\boldsymbol{\eta}_d = \mathbf{X}_d\boldsymbol{\beta} + \mathbf{Z}_d\mathbf{V}$ in the same way: $\begin{bmatrix} \boldsymbol{\eta}_{ds} \\ \boldsymbol{\eta}_{dr} \end{bmatrix} = \begin{bmatrix} \mathbf{X}_{ds} \\ \mathbf{X}_{dr} \end{bmatrix}\boldsymbol{\beta} + \begin{bmatrix} \mathbf{Z}_{ds} \\ \mathbf{Z}_{dr} \end{bmatrix}\mathbf{V} = \begin{bmatrix} \mathbf{X}_{ds}\boldsymbol{\beta} + \mathbf{Z}_{ds}\mathbf{V} \\ \mathbf{X}_{dr}\boldsymbol{\beta} + \mathbf{Z}_{dr}\mathbf{V} \end{bmatrix}$. We use $\hat{\mathbf{y}}_{N_d} = \sum_{i_s=1}^{n_d}\mathbf{y}_{di_s} + \sum_{i_r=1}^{N_d-n_d}\hat{\boldsymbol{\mu}}_{di_r}$ as a predictor of $\mathbf{y}_{N_d}$, where $\hat{\boldsymbol{\mu}}_{di_r} = \underset{1\le k\le K-1}{col}(\hat{\mu}_{di_rk})$ with $\hat{\mu}_{di_rk} = \frac{\exp(\mathbf{x}_{di_r}\hat{\boldsymbol{\beta}}_k + \hat{v}_{dk})}{1+\sum_{k=1}^{K-1}\exp(\mathbf{x}_{di_r}\hat{\boldsymbol{\beta}}_k + \hat{v}_{dk})}$ for crop $k = 1, 2, \cdots, K-1$ and $\hat{\mu}_{di_rK} = \frac{1}{1+\sum_{k=1}^{K-1}\exp(\mathbf{x}_{di_r}\hat{\boldsymbol{\beta}}_k + \hat{v}_{dk})}$ for crop K. Here, $\hat{\boldsymbol{\beta}}_k$ is the $\boldsymbol{\beta}_k$ estimate at the convergence iteration $\hat{\boldsymbol{\beta}}^{(m)} = \underset{1\le k\le K-1}{col}\left(\hat{\boldsymbol{\beta}}_k^{(m)}\right)$ and $\hat{v}_{dk}$ is the predictor of the small-area random effect v_{dk} in $\mathbf{v} = \underset{1\le d\le D}{col}\ \underset{1\le k\le K-1}{col}(v_{dk})$ defined by $\hat{\mathbf{v}} = \underset{1\le d\le D}{col}\ \underset{1\le k\le K-1}{col}(\hat{v}_{dk}) = \hat{\boldsymbol{\Sigma}}\mathbf{Z}^T\hat{\mathbf{V}}\boldsymbol{\xi}^{-1}\left(\boldsymbol{\xi} - \mathbf{X}\hat{\boldsymbol{\beta}}\right)$, where $\hat{\boldsymbol{\Sigma}} = \hat{\boldsymbol{\theta}} \otimes \mathbf{I}_\mathbf{D}$ and $\hat{\boldsymbol{\theta}}$ is the $\boldsymbol{\theta}$ estimate at the convergence iteration $\hat{\boldsymbol{\theta}}^{(m)} = \underset{1\le k\le K-1}{row}\ \underset{1\le k'\le K-1}{col}\left(\hat{\sigma}_{kk'}^{(m)2}\right)$.

B.3 Mean squared error

The mean squared error of predictor $\hat{\mathbf{y}}_{N_d}$ is $MSE\left(\hat{\mathbf{y}}_{N_d}\right) = MSE(\hat{\boldsymbol{\tau}}_{dr}) + \sum_{i_r=1}^{N_d-n_d}\mathbf{V}\mathbf{y}_{di_r}$. The term $MSE(\hat{\boldsymbol{\tau}}_{dr})$ can be approximated by a Taylor series expansion of $\hat{\boldsymbol{\mu}}_{di} = h(\hat{\boldsymbol{\eta}}_{di})$ around $\boldsymbol{\mu}_{di} = h(\boldsymbol{\eta}_{di})$, where $\boldsymbol{\eta}_{di} = \mathbf{X}_{di}\boldsymbol{\beta} + \mathbf{Z}_{di}\mathbf{V}$.

As a result, we have $\hat{\boldsymbol{\mu}}_{di} - \boldsymbol{\mu}_{di} \simeq \mathbf{H}_{di}(\hat{\boldsymbol{\eta}}_{di} - \boldsymbol{\eta}_{di})$. Let $\boldsymbol{\tau}_{dr}^T = \sum_{i_r=1}^{N_d-n_d} \mathbf{H}_{di_r}\boldsymbol{\eta}_{di_r} = \sum_{i_r=1}^{N_d-n_d} (\mathbf{H}_{di_r}\mathbf{X}_{di_r}\boldsymbol{\beta} + \mathbf{H}_{di_r}\mathbf{Z}_{di_r}\mathbf{V}) = \mathrm{M}_{dr}\boldsymbol{\beta} + \mathrm{K}_{dr}\mathbf{V}$, where $\mathrm{M}_{dr} = \sum_{i_r=1}^{N_d-n_d} \mathbf{H}_{di_r}\mathbf{X}_{di_r}$ and $\mathrm{K}_{dr} = \sum_{i_r=1}^{N_d-n_d} \mathbf{H}_{di_r}\mathbf{Z}_{di_r}$. The approximation of $MSE(\hat{\boldsymbol{\tau}}_{dr}) = MSE(\hat{\boldsymbol{\tau}}_{dr}^T)$ is $MSE(\hat{\boldsymbol{\tau}}_{dr}) = \mathbf{G}_{1d}(\boldsymbol{\theta}) + \mathbf{G}_{2d}(\boldsymbol{\theta}) + \mathbf{G}_{3d}(\boldsymbol{\theta}) + \mathbf{G}_{4d}(\boldsymbol{\theta})$, where $\mathbf{G}_{1d}(\boldsymbol{\theta}) = \mathrm{K}_d\mathrm{T}\mathrm{K}_d^T$ $\mathbf{G}_{2d}(\boldsymbol{\theta}) = \Lambda_d\mathrm{P}\Lambda_d^T$, $\mathbf{G}_{3d}(\boldsymbol{\theta}) = \left(\frac{\partial\Gamma_d}{\partial\theta}\right)^T \mathbf{V}\left(\frac{\partial\Gamma_d}{\partial\theta}\right)\mathbf{V}_{\boldsymbol{\theta}}^{-1}$ and $\mathbf{G}_{4d}(\boldsymbol{\theta}) = \mathbf{V}\mathbf{y}_{dr} = \sum_{i_r=1}^{N_d-n_d} \mathbf{V}\mathbf{y}_{di_r}$, with $\mathrm{T} = \left(\mathbf{Z}^T\mathrm{H}\mathrm{Z} + \boldsymbol{\Sigma}^{-1}\right)^{-1}$, $\mathbf{H} = \underset{1 \le d \le D}{diag}\ \underset{1 \le i \le N_d}{diag}\ (\mathbf{H}_{di})$, $\mathrm{P} = \mathbf{V}\hat{\boldsymbol{\beta}} = \left(\mathbf{X}^T\mathbf{V}^{-1}\mathbf{X}\right)^{-1}$, $\mathbf{V} = \mathrm{Z}\Sigma\mathrm{Z}^T + \mathbf{Ve}$, $\Lambda_d = \mathbf{m}_d - \mathrm{K}_d\mathrm{T}\mathrm{Z}^T\mathbf{VeX}$, and $\frac{\partial\Gamma_d}{\partial\theta} = \mathbf{V}^{-1}\mathrm{Z}\mathrm{K}_d^T$. Here, $\mathbf{V}_{\boldsymbol{\theta}}$ denotes the Fisher information of the variance component parameters $\boldsymbol{\theta}$.

As estimator of the mean squared error of predictor $\hat{\mathbf{y}}_{N_d}$, we use the asymptotically unbiased estimator proposed in [23], $\hat{MSE}\left(\hat{\mathbf{y}}_{N_d}\right) = \mathbf{G}_{1d}\left(\hat{\boldsymbol{\theta}}\right) + \mathbf{G}_{2d}\left(\hat{\boldsymbol{\theta}}\right) + 2\mathbf{G}_{3d}\left(\hat{\boldsymbol{\theta}}\right) + \mathbf{G}_{4d}\left(\hat{\boldsymbol{\theta}}\right)$, where $\hat{\boldsymbol{\theta}}$ is the $\boldsymbol{\theta}$ estimate at the convergence iteration.

Author details

Luis Ambrosio[1*], Luis Iglesias[1], Carmen Marin[1], Sophie Bontemps[2], Böris Nörgaard[2], Pierre Houdmont[2], Cosmin Cara[3], Cosmin Udroin[3], Laurentiu Nicola[3], Sadri Haouet[4] and Zoltan Szantoi[5,6]

1 Technical University of Madrid, Madrid, Spain

2 Catholic University of Louvain, Louvain, Belgium

3 CS Group Romania, Craiova, Romania

4 CLS, Lille, France

5 European Space Agency, Frascati, Italy

6 Stellenbosch University, Stellenbosh, South Africa

*Address all correspondence to: luis.ambrosio@upm.es

References

[1] Ambrosio L, Iglesias L, Marin C, Deffense N. Integration of remote sensing data into national statistical office sampling designs for agriculture. Statistical Journal of the IAOS. 2023; **39**(2):473-489. DOI: 10.3233/SJI-220116

[2] Sawin PH, Swain PH, Davis SM. editors. Remote Sensing: The Quantitaive Approach, pp136-187. New York: McGraw-Hill; 1978, 396 p

[3] Golan A, Judge G, Miller D. Maximum entropy econometrics. New York: Wiley; 1996. 307 p

[4] FAO. Handbook on master sampling frames for agricultural statistics. Frame development, sample design and estimation. In: Improving Agricultural and Rural Statistics. Global Strategy. Rome: FAO; 2015. 170 p

[5] Cotter J, Nealon J. Area frame design for agricultural surveys. Washington, D. C.: Research and Applications Division. National Agricultural Statistics Service. United States Department of Agriculture; 1987

[6] FAO. Multiple Frame Agricultural Surveys. Volume 1: Current Surveys Based on Area and List Sampling Methods. Volume 2: Agricultural Survey Programmes Based on Area Frame or Dual Frame (Area and List) Sample Designs, Statistical Development Series 7 and 10. Roma: FAO; 1998

[7] Gay C, Porchier JC. Land cover and land use classification using TER-UTI. In: Holland TE and Van den Broecke MPR, Editors. Proceedings of Agricultural Statistics 2000, International Statistical Institute. Voorburg, The Netherlands. 1998. pp. 193-201. Available from: https://www.isi-web.org/isi.cbs.nl/iamamember/Books/agric2000/page-193.pdf

[8] Nusser SM, Goebel JJ. The national resources inventory: A long-term multi-resource monitoring programme. Environmental and Ecological Statistics. 1997;**4**(3):181-204. DOI: 10.1023/A:1018574412308

[9] Nusser SM, Goebel JJ, Fuller WA. Design and estimation for investigating the dynamics of natural resources. Ecological Applications. 1998;**8**(2):234-245. DOI: 10.1890/1051-0761(1998)008[0234:DAEFIT]2.0.CO;2

[10] Directorate of Forecasting Analysis and Agricultural Statistics. Ministry of Agriculture and Rural Equipment. Senegal: Senegal-Annual Agricultural Survey. Methodology and sampling plan of the agricultural survey; 2017. Available from: http://anads.ansd.sn/index.php/catalog/218/download/1846

[11] Fuller WA. Sampling Statistics. New York: Wiley; 2009. 472 p. DOI: 10.1002/9780470523551

[12] Agresti A. Categorical data analysis. New York: Wiley; 2002. 710 p

[13] McCulloch CE, Searle SR. Generalized, Linear, and Mixed Models. New York: Wiley; 2001, 325 p

[14] Hogland J, Billor N, Anderson N. Comparison of standard maximum likelihood classification and polytomous logistic regression used in remote sensing. European Journal of Remote Sensing. 2013;**46**(1):623-640. DOI: 10.5721/EuJRS20134637

[15] Ambrosio L, Iglesias L. Land cover estimation in small areas using ground surveys and remote sensing. Remote

Sensing of Environment. 2000;**74**: 240-248. DOI: 10.1016/S0034-4257(00)00114-0

[16] Cressie N. Statistics for Spatial Data. New York: John Wiley & Sons; 1991. 900 p

[17] Ambrosio L, Iglesias L, Marin C. Systematic sample design for the estimation of spatial means. Environmetrics. 2003;**14**(1):45-61. DOI: 10.1002/env.564

[18] Fuller WA, Isaki CT. Survey design under superpolutation models. In: Krewski D, Rao JNK, Platek R, editors. Current Topics in Survey Sampling. New York: Academic Press; 1981. pp. 199-226

[19] Ambrosio L, Iglesias L. Identifying the most Appropriate Sampling Frame for Specific Landscape Types, Technical Report Series. GO-01-2014. FAO. Available from: https://www.fao.org/3/ca6436en/ca6436en.pdf

[20] Ambrosio L, Iglesias L, Marin C. A model-assisted approach to identify a cost-efficient spatial sampling strategy. In: Paper Presented at: Eighth International Conference on Agricultural Statistics (ICAS-VIII). New Delhi (IN); 2019

[21] Ypma J, Johnson SG, Stamm A. lnoptr R package [software]. Version 2.0.3.9100. Available from: https://astamm.github.io/nloptr/index.html

[22] Saei A, Chambers R. Small Area Estimation under Linear and Generalized Linear Mixed Models with Time and Area Effects. Methodology Working Paper M03/15. Southampton (UK): Southampton Statistical Sciences Research Institute; 2003: 31 p. http://eprints.soton.ac.uk/id/eprint/8165

[23] Molina I, Saei A, Lombardia MJ. Small area estimates of labour force participation under a multinomial logit mixed model. Journal of the Royal Statistical Society. Series A, Statistics in Society. 2007;**170**(4):975-1000. DOI: 10.1111/j.1467-985X.2007.00493.x

[24] Lopez-Vizcaino E, Lombardia MJ, Morales D. Small area estimation of labour force indicators under a multinomial model with correlated time and area effects. Journal of the Royal Statistical Society. Series A, Statistics in Society. 2015;**178**(3):535-565. DOI: 10.1111/rssa.12085

Chapter 7

Evaluation of Morphometric Parameters for Identifying Groundwater Recharge Potential Zone by GIS Application in Kufrah Basin SE Libya

Jamal Zamot and Mohammed Afkareen

Abstract

A number of studies have been done by using various approaches of remote sensing techniques and GIS applications to discover and identify a great aquifer in SE Libya. One of the main goals of these studies was to find out the source water of this aquifer in the Kufrah basin and how it could be recharged. Modern GIS applications were used in the current study for morphometric analysis of the network streams and delineation of Kufrah basin to identifying groundwater recharge zone. The obtained results show a positive linear relation. The results show that the study area is a flat area with low relief topography, low moisture content and high evaporation rate. The drainage pattern of the Kufrah basin is a dendritic pattern with the 7th order of streams. The network streams over a permeable subsurface material, high infiltration capacity with very coarse texture, and low runoff. The Kufrah river can be described as an "Old age" river and drained inland at Al-Jaghbub area.

Keywords: morphometric analysis, GIS, aquifer, hypsometric curve, Kufrah basin

1. Introduction

In Libya, the overall usage of water accounts for 97% of groundwater, with less than 3% coming from other sources. A part of the Nubian sandstone aquifer, which is an integral part of the biggest fossil aquifer system in Africa, is located in Libya to the east of the Sahara Desert [1, 2]. This enormous aquifer is believed to have recharged during the Pleistocene and Pliocene epochs, which are known for their extended periods of rainy weather in Libya's Quaternary history [3, 4]. The presence of freshwater fauna in the Sahabi's paleochannel, in addition to the width and depth of the Sahabi Canyon, strongly suggests that a massive river drained the southern part of Libya to the Mediterranean Sea *via* the Sahabi paleochannel and, more recently, internally into the Al-Jaghbub depression's inland delta [5, 6]. In semiarid and desert countries like Libya, recharging of reservoirs is critical for dealing with population growth, as well as developing water supplies [7].

Although the Nubian sandstone aquifer in Libya is estimated to supply 1020.7 million cubic meters of water per year for all water purposes, it could also be used sustainably [1, 2]. In order to avoid groundwater depletion, the determination of the groundwater recharge zone is essential. There is a direct correlation between the geomorphic parameters, soil profile characteristics, and geological properties of groundwater recharge. Rainfall, drainage density, texture ratio, circularity and elongation ratio, infiltration rate, topography, lithology, and geomorphology all influence where and how groundwater occurs and travels.

However, morphometric analysis of network streams and hydrologic network delineation are thought to be successfully accomplished with the use of remote sensing and GIS applications, which offer a substitute or supplement to field observations and numerical models.

Moreover, remote sensing and GIS techniques have proofed their abilities in penetrating the ground surface and distinguishing the buried features such as channel system (e.g., [8–13]). Therefore, these tools were significantly used in this study for identifying, discovering, and mapping the water shed of Kufrah basin in Libya. In the current study, GIS application with conventional mathematical equations were utilized for morphometric analysis and delineation of the Kufrah basin in SE of Libya.

Accordingly, in order to approximate the level of development of the drainage network of the Kufrah basin, a hypsometric curve was made to represent the distribution of elevations throughout the river's catchment region.

2. Location of the study area

The study area is named Kufrah basin, which is located in southeastern part of Libya and covers an area of about 175,000 km^2. It is bordered by Egypt to the east,

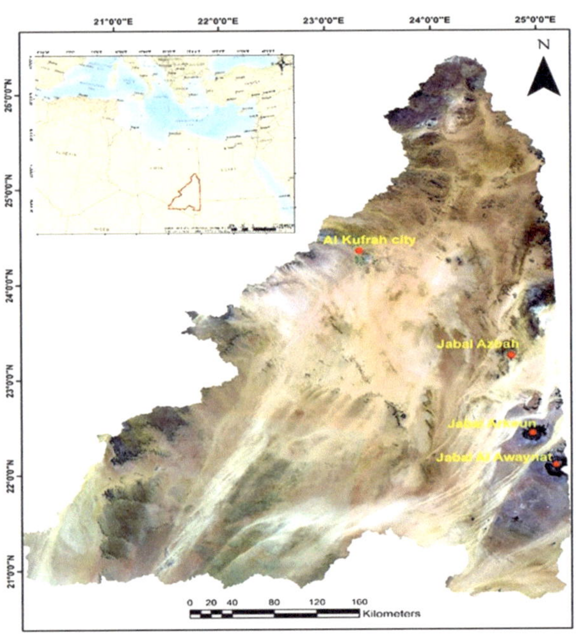

Figure 1.
Shows the location map of the study area, SE Libya.

Sudan and Chad to the south, and high Tibesti Volcanic Province to the west and from the north by the Al-Jaghbub depression, **Figure 1**.

3. Data and methodology

The main data are a Digital Elevation Model (DEM) over the study area that was taken from the Shuttle Radar Topography Mission (SRTM), with a spatial resolution equal to 90 meters. In addition to the geological map of Kufrah basin from Industrial Research Centre (IRC) with 1:250,000 scale [14].

In order to visualize the digital data, ArcMap software was used. The morphometric analysis, hydrology drainage network, and basin delineation have been obtained from DEM by using ArcGIS hydrology tools (10.5), besides a manually mathematical analysis by using equations.

4. The geology of the area

The interested aquifer is located within Kufrah basin, which is an intracratonic sag basin located in SE Libya [15, 16]. The sides of the basin are bordered by highs that made up of the basement, where a succession of Proterozoic and Paleozoic strata crop out, Tibesti Massif in the west, Jebel Uweinat in the east, the Ennedi and Borkou in the south, and Jebel Dalma in the north, **Figure 2** [15, 16].

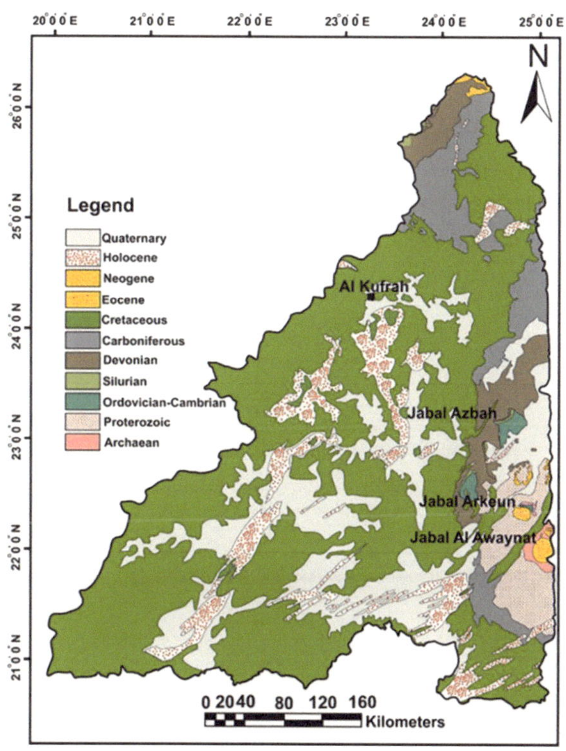

Figure 2.
The geological map of the Kufrah basin.

The Kufrah basin is generally filled with Paleozoic and Mesozoic sedimentary strata. Fluvial, shallow marine sandstones, and shales are the most characterization of the deposition in Cambrian-Ordovician time. The strata of Triassic to Early Cretaceous are made of continental sandstones (Nubian Sandstone) that rest uncomfortably over Paleozoic rocks. The Mesozoic sediments are overlain by fluvial deposits of Messinian to Pliocene and Quaternary eolian deposits (**Figure 2**) [15, 17–19].

5. Results and discussion

The results of morphometric analysis of Kufrah basin are listed in **Tables 1** and **2**, respectively, and they are under major heads of slope, areal, and relief aspects.

5.1 Slope, aspect, and relative relief parameters

One of the most important parameters in morphometric analysis is the slope parameter, which determines how rainwater will run off along the network of the

Morphometric parameters	Formula	Results	Reference
Basin length (Lb)	GIS Software	733 km	[20]
Basin perimeter (P)	GIS Software	3332 km	[20]
Basin area (A)	GIS Software	174,591 km^2	[20]
Mean Basin width (Wb)	*Wb = A / Lb; where An area of watershed; Lb basin length.*	238.18	[21]
Drainage density (Dd)	*Dd = Lu / A; where Lu total length of streams; An area of watershed*	0.18 km^2	[21]
Drainage intensity (Di)	*Di = Fs / Dd; where Fs Stream frequency; Dd Drainage density*	0.14	[22]
Drainage texture (Dt)	*Dt = Nu / P; where Nu total number of streams; P Basin perimeter.*	1.28	[23]
Stream frequency (Sf)	*Sf = Nu / A; where Nu total number of streams; A area of watershed*	0.0243 km^2	[21]
Texture ratio (Rt)	*Rt = ΣN1/P, where N1 the total number of streams of 1st order, P perimeter of watershed.*	0.004	[20]
Form factor (Rf)	$Rf = A/(L)^2$*; Where A area of watershed; L basin length.*	0.32	[21]
Circularity ratio (Rc)	$Rc = 4\pi A/P^2$; $\pi = 3.14$*, A area of watershed, P perimeter of watershed.*	0.197	[24]
Elongation ratio (Re)	$Re = 2\sqrt{(A/\pi)}/L$*; where* $\pi = 3.14$*, An area of watershed, L length of watershed.*	0.0209	[20]
Length of overland flow (Lof)	*Lof = 1/2Dd; Dd Drainage density*	2.8 mature stage	[23]
Infiltration number (In)	*In = Dd * Sf; Sf Stream frequency, Dd Drainage density.*	0.0044	[22]
Constant channel maintenance (Ccm)	*Ccm = 1 / Dd; where Dd Drainage density*	5.5	[20]

Table 1.
List of the morphometric parameters and the formulas used in Kufrah Paleo river analysis.

Relief parameters	Formula	Results	Reference
Maximum elevation in the area (h_{Max})	_	1887 m	_
Minimum elevation in the area (h_{Min})	_	239 m	_
Basin relief (H)	$H = h_{Max} - h_{Min}$	1648 m	[20]
Relief ratio (Rf)	*Rf = H/L; where H basin relief; L length of basin.*	2.25	[20]
Ruggedness index (RI).	*Rn = H* Dd; where H basin relief, Dd drainage density.*	0.29 high erosion	[25]
Dissection index (DI)	*DI = H/* h_{Max}*; where H basin relief,* h_{Max} *maximum elevation in the area.*	0.873	[20]

Table 2.
List of the relief parameters and the formulas used in Kufrah Basin analysis.

river basin under the effecting of slope [26]. The slope degree map, elevation map, and the relative relief map have shown the study area as a flat area with low relief topography that makes the runoff slow, **Figure 3**. Consequently, this will give the water more time to seep into the subsurface, greatly recharging the groundwater. However, the highest slope in the study area is only noticed on the far east, where Jebel Uweinat and Jabal Arkenu are located.

The Kufrah water basin starts from an of elevation 1887 m in the south and terminated in the inland delta at Al-Jaghbub depression in the north with 239 m. It has a dendritic drainage pattern, which demonstrates that the rocks and soil type are uniformly homogeneous.

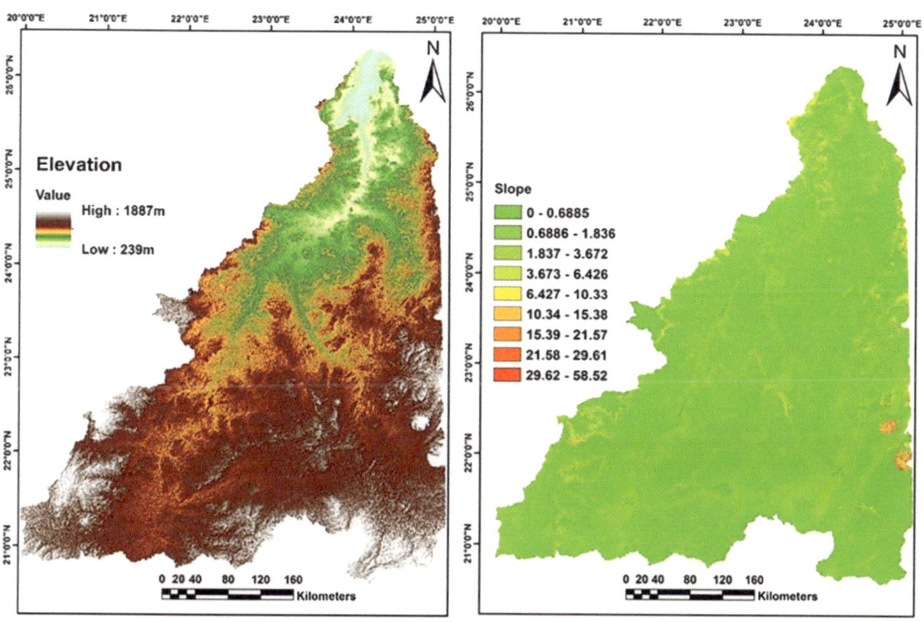

Figure 3.
Illustration of the elevation map and the slope map of the Kufrah basin.

The direction that a sloped landscape faces is referred to as its aspect. The research area's vegetation distribution, biodiversity, and precipitation patterns can all be significantly impacted by the slope aspect [27, 28]. From the value of the raster data output, the aspect's compass direction was determined. The aspect map of Kufrah basin is mainly dominated by flat to very gentle aspect slope, **Figure 4**.

The hillshade map is highlighted the structural effect of Jebel Awaynat and Jabal Arkenu on the eastern flank of the basin, while the rest of the region is not **Figure 4**.

5.2 Aerial morphometric aspect

The morphometric parameters of drainage density (Dd), drainage texture (T), texture ratio (Rt), infiltration number (In), form factor (Rf), stream frequency (Fs), elongation ratio (Re), circulatory ratio (Rc), and length of overland flow (Lof) are included in the aerial morphometric aspects. Climate conditions, geological structure, and lithology all play a role in determining the drainage basin's characteristics [29]. Kufrah water basin is believed to be developed over non-resistance rocks, and thereby, the number and order of its streams is high with the 7th order of streams, **Figure 5**.

5.2.1 Drainage density (Dd)

Horton [21] initially proposed drainage density (Dd) as a crucial measure of the linear scale of the landform element in the stream-eroded terrain. He defined drainage density as the ratio of the total length of the stream in a given drainage basin and the area of that drainage basin. Furthermore, a variety of characteristics, including the subsurface material, relief, surface runoff, infiltration capacity, flood volumes, and density of vegetation, contribute to both high and low drainage density values.

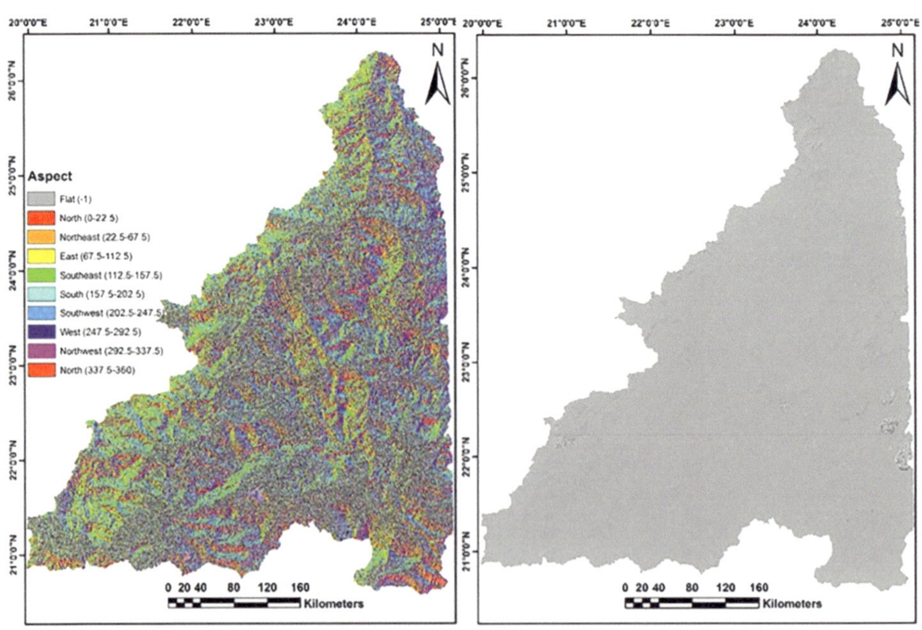

Figure 4.
The aspect map and the hillshade map of the Kufrah basin.

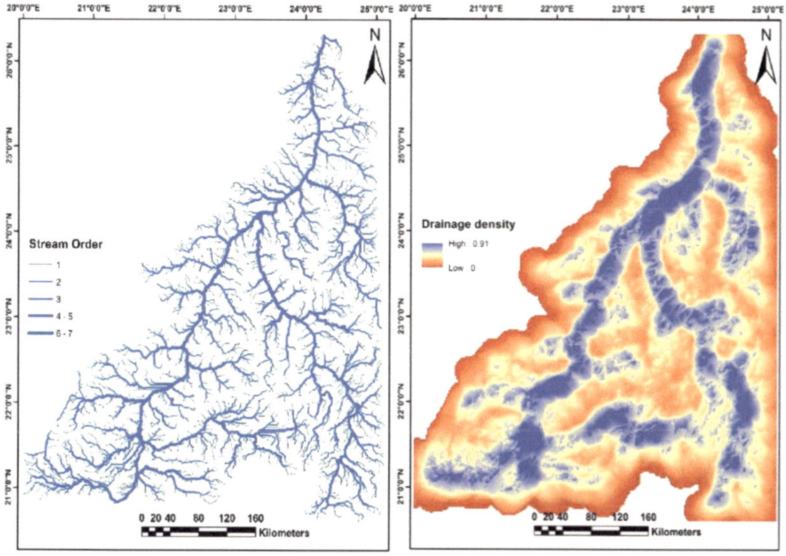

Figure 5.
The stream orders and the drainage density of Kufrah basin.

The drainage density of the study basin is 0.18 km^2 that indicates high permeable subsurface and coarse drainage, **Figure 5**.

5.2.2 Drainage texture (T)

Drainage texture in the present study shows the value of Kufrah watershed with 1.28, which indicates a texture pattern to be a coarse drainage texture.

5.2.3 Texture ratio (Rt)

The study area has a very low texture ratio of 0.004, which indicates permeable subsurface materials, low relief conditions, and high infiltration capacity with very coarse drainage texture.

5.2.4 Infiltration number (In)

The infiltration number of the basin is 0.0044, indicating high infiltration materials and low surface runoff.

5.2.5 Form factor (Rf)

The basin form of the Kufrah basin is slightly elongated, as indicated by its form factor value of 0.32, which also indicates that the basin has low peak flow.

5.2.6 Stream frequency (Fs)

The high value of stream frequency indicates more surface runoff and vice versa. In the present study, the stream frequency value of the basin is low, which indicates more groundwater potential.

5.2.7 Elongation ratio (Re)

The elongation value of the water basin is 0.0209. The low value indicates that the basin has low relief with moderate to high infiltration capacity and low runoff.

5.2.8 Circularity ratio (Rc)

A circular basin is more efficient in runoff discharge than an elongated basin. Circularity ratio is 0.197, which indicates less elongated basin with low relief and high infiltration capacity.

5.2.9 Length of overland flow (Lof)

The length of the overland flow of the study river is 2.8, and the value is slightly high, indicating that the basin is consists of low relief and slope, and the river is in the mature stage.

5.3 Relief aspects

The relief aspects include Dissection index (Di), Ruggedness Index (RI), and Relief Ratio (Rf). These aspects are calculated and presented in **Table 2**.

5.3.1 Dissection index

According to Singh [30], the Dissection index (DI) is an important morphometric indicator of nature and magnitude of dissection of terrain, which is expressing the ratio of the maximum relative relief to a maximum absolute relief. The DI value is between zero, which indicates complete absence of dissection or vertical erosion and one that reveals vertical cliff. Generally, the areas with high DI indicate high relative relief where the slope of the land is steep and unstable that results in enhanced erosion. On the contrary, low DI corresponds with low relative relief and with the subdued relief or old stage where the land is flat and more stable [31]. The DI of the study area is 0.873, which indicates the basin is a moderately to highly dissected, **Figure 6**.

5.3.2 Ruggedness Index (RI)

The ruggedness number of the basin is 0.29, which indicates higher soil erosion susceptibility, **Figure 6**.

5.3.3 Relief ratio (Rf)

The relief ratio (Rh) of maximum relief to horizontal distance along the longest dimension of the basin parallel to the principal drainage line is termed as a relief ratio [20]. The relief ratio of the river basin is 2.25, indicating that this basin is composed of nonresistant rocks, and the basin is under low relief and gentle slope.

5.4 Bifurcation ratio

The bifurcation ratio (Rb) can be defined as a ratio of the number of stream branches of given order to the number of branches of next higher order [32]. By

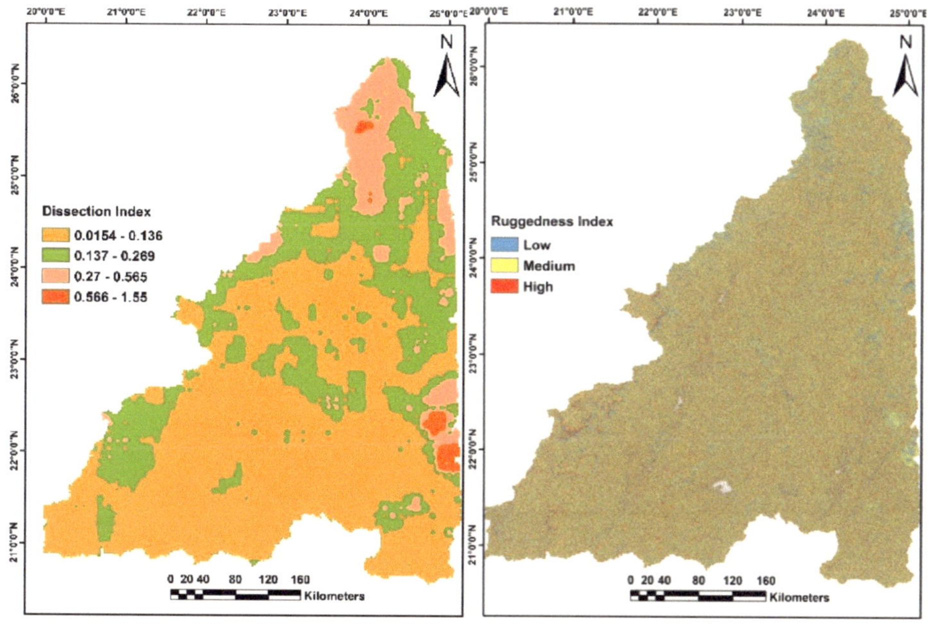

Figure 6.
The dissection index map and the ruggedness index map of Kufrah basin.

examining the irregularities in the bifurcation ratio, the bifurcation analysis can provide information about the drainage pattern's structural disturbances, geological and lithological development of the drainage basin, and drainage integration [33]. The bifurcation ratio of the Kufrah watershed is given in **Table 3**, where the bifurcation ratio is (1.8) that reveals that the area does not structurally controlled with flat terrains.

5.5 Valley stage

Finally, a hypsometric curve for the Kufrah basin was made in order to surmise the drainage network's stage of development. The hypsometric curve of a catchment represents the relative area below (or above) a given altitude [34]. The shape of the hypsometric curve provides valuable information on the erosional stage of the basin and also on the tectonic factors controlling it [35].

Rb = Nu/Nu + 1, where Nu = total number of stream segments of order; Nu +1 = number of segments of next higher order	**Bifurcation ratios (Rb)**
1st/2nd	2.09
2nd/3rd	1.7
3rd/4th	2.8
4th/5th	1.9
5th/6th	0.8
6th/7th	1.8

Table 3.
The bifurcation ratios (Rb) of streams order of Kufrah basin.

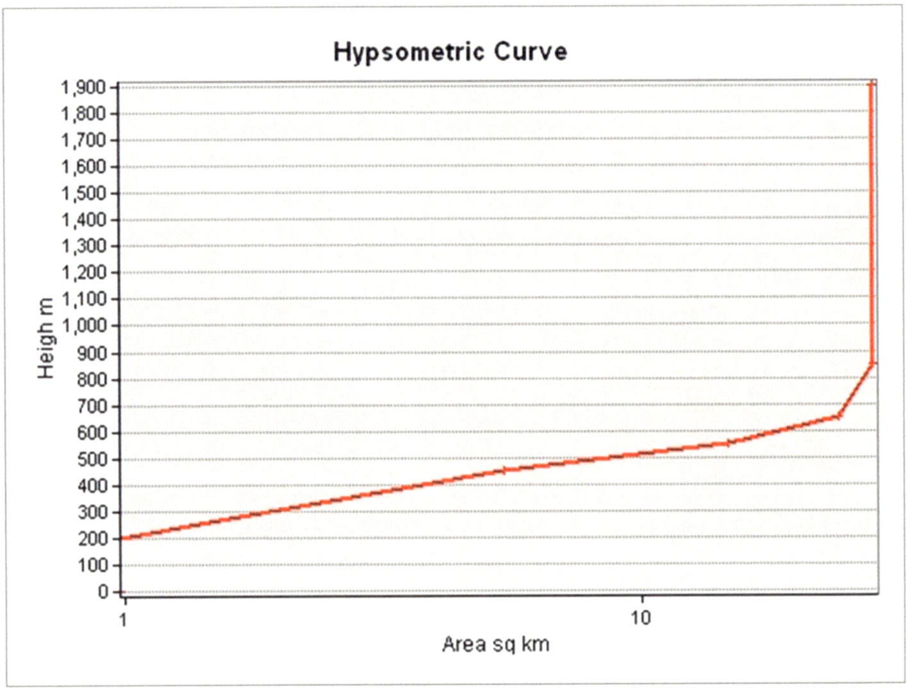

Figure 7.
A hypsometric curve for Kufrah water basin.

However, rivers can be classified according to specific characteristics as initial a "Youthful", "Mature," and "Old Age". From **Figures 3** and **4**, the Kufrah network river can be described as an "Old age" river, showing a more S-shaped feature and displaying a concave upward feature at higher elevation and concave downward at lower elevations, which characterizes an old age of drainage basin, **Figure** 7.

6. Conclusion

The present study has used GIS applications and remote sensing techniques for morphometric analysis and stream delineation of Kufrah basin. These techniques have demonstrated their ability to penetrate the ground surface and identify the morphometry of the buried streams. All the morphometric parameters that have been analyzed indicated same results. The used tools have revealed that Kufrah water basin was running from the south to the north and drained internally into the Al-Jaghbub depression. The Kufrah basin has a very coarse drainage texture, high infiltration capacity, high permeability of subsurface material, and low surface runoff. Furthermore, the basin is composed of nonresistant rocks and thus higher soil erosion susceptibility. The water basin is a moderately to highly dissected with low structural effects. Finally, the drainage network in Kufrah basin is in an "Old age stage."

Conflict of interest

The authors declare no conflict of interest.

Author details

Jamal Zamot[1]* and Mohammed Afkareen[2]

1 Earth Sciences Department, Benghazi University, Benghazi, Libya

2 Sirte Oil Company, Marsa El-Brega, Libya

*Address all correspondence to: jamal.zamot@uob.edu.ly

References

[1] Hamad S, Saaid F. Characterization and management evaluation of the nubian sandstone aquifer in Tazerbo wellfield of the Libyan man-made river project. Applied Water Science. 2022;**12**(7):165

[2] Ahmed M. Sustainable management scenarios for northern Africa's fossil aquifer systems. Journal of Hydrology. 2020;**589**:125196

[3] Ghoneim E, Benedetti M, El-Baz F. An integrated remote sensing and GIS analysis of the Kufrah Paleoriver, eastern Sahara. Geomorphology. 2012;**139**:242-257

[4] De Menocal P, Ortiz J, Guilderson T, Adkins J, Sarnthein M, Baker L, et al. Abrupt onset and termination of the African humid period: Rapid climate responses to gradual insolation forcing. Quaternary Science Reviews. 2000;**19**(1-5):347-361

[5] Hecht MK. Fossil snakes and crocodilians from the Sahabi formation of Libya. In: Boaz NT, El-Arnauti A, Gaziry AW, de Heinzelin J, Boaz, DD, editors, Neogene Paleontology and Geology of Sahabi, Libya. Alan R. Liss, New York; 1987. pp. 101-106

[6] Boaz NT. Sahabi and Neogene hominoid evolution. In: Boaz NT, El-Arnauti A, Gaziry AW, de Heinzelin J, Boaz DD, editors, Neogene Paleontology and Geology of Sahabi, Libya. Alan R. Liss, New York; 1987. pp. 129-134

[7] Kassa AK, Tessema N, Habtamu A, Girma B, Adane Z. Identifying groundwater recharge potential zone using analytical hierarchy process (AHP) in the semi-arid Shinile watershed, Eastern Ethiopia. Water Practice & Technology. 2023;**18**(11):2834-2850

[8] Roth LE, Elachi C. Coherent electromagnetic losses by scattering from volume inhomogeneities. IEEE Transactions on Antennas and Propagation. 1975;**23**:674-675

[9] McCauley JF, Schaber GG, Breed CS, Grolier MJ, Haynes CV, Issawi B, et al. Subsurface valleys and geoarchaeology of the eastern Sahara revealed by shuttle radar. Science. 1982;**218**:1004-1020

[10] El-Baz F, Mainguet M, Robinson CA. Fluvio-aeolian dynamics in the North-Eastern Sahara: Interrelation between fluvial and aeolian systems and implications to ground water. Journal of Arid Environments. 2000;**44**:173-183

[11] Robinson CA, El-Baz F, Al-Saud TSM, Jeon SB. Use of radar data to delineate palaeodrainage leading to the Kufra oasis in the eastern Sahara. Journal of African Earth Sciences. 2006;**44**:229-240

[12] Ghoneim E, El-Baz F. The application of radar topographic data to mapping of a mega-paleodrainage in the Eastern Sahara. Journal of Arid Environments. 2007a;**69**:658-675

[13] Paillou A, Schuster M, Tooth T, Farr T, Rosenqvist A, Lopez S, et al. Mapping of a major paleodrainage system in eastern Libya using orbital imaging radar: The Kufrah River. Earth and Planetary Science Letters. 2009;**277**:327-333

[14] Saïd MM, Oweiss KA, Mehidi BO. Jabal Arkenu sheet: Explanatory booklet. In: Joint Geological Mapping Project between the Egyptian Geological Survey and Mining Authority (EGSMA) and the Industrial Research Centre (IRC) of Libya. Cairo; 2000

[15] Lüning S, Craig J, Fitches B, Mayouf J, Busrewil A, El Dieb M, et al. Re-evaluation of the petroleum potential of the Kufra Basin (SE Libya, NE Chad): Does the source rock barrier fall? Marine and Petroleum Geology. 1999;**16**:693-718

[16] Hallett D. Petroleum Geology of Libya. Amsterdam: Elsevier; 2002

[17] Bellini E, Massa D. A stratigraphic contribution to the Palaeozoic of the southern basins of Libya. In: Salem NJ, Busrewil MT, editors. The Geology of Libya. Vol. 1. Tripoli: Academic Press; 1980. pp. 1-56

[18] Bellini E, Giori I, Ashuri O, Benelli F. Geology of Al Kufrah Basin, Libya. In Symposium on the geology of Libya. 1991. pp. 2155-2184

[19] Gindre L, Le Heron D, Bjørnseth HM. High resolution facies analysis and sequence stratigraphy of the Siluro-Devonian succession of Al Kufrah basin (SE Libya). Journal of African Earth Sciences. 2012;**76**:8-26

[20] Schumm SA. Evolution of drainage systems and slopes in badlands at Perth Amboy, New Jersey. Geological Society of America Bulletin. 1956;**67**:597-646. DOI: 10.1130/0016-7606(1956)67[597:EODSAS]2.0.CO;2

[21] Horton RE. Drainage basin characteristics. American Geophysical Union Transactions. 1932;**13**:348-352

[22] Faniran A. The index of drainage intensity: A provisional new drainage factor. The Australian Journal of Science. 1968;**31**(9):326-330

[23] Horton RE. Erosional development of streams and their drainage basins: Hydro-physical approach to quantitative morphology. Geological Society of American Bulletin. 1945;**56**:275-370

[24] Miller VC. A quantitative geomorphic study of drainage basin characteristics in the Clinch Mountain area, Virginia and Tennessee. New York: Columbia University; 1953:**3**

[25] Melton MA. An Analysis of the Relations among Elements of Climate, Surface Properties, and Geomorphology. Vol. 11. New York: Department of Geology, Columbia University; 1957

[26] Villela SM, Mattos A. Hidrologia aplicada. Editora McGraw-Hill do Brasil; 1975

[27] Zamot J, Afkareen M. Geomorphological parameters by remote sensing and GIS techniques (A case study of flash flood in Mikhili Village, Al Jabal Al Akhdar, NE of Libya). In: The Forth International Conference for Geospatial Technologies – Libya GeoTec. Vol. 4. 2020

[28] Khakhlari M, Nandy A. Morphometric analysis of Barapani river basin in Karbi Anglong District, Assam. International Journal of Scientific and Research Publications. 2016;**6**(10):238-249

[29] Resmi MR, Babeesh C, Hema A. Quantitative analysis of the drainage and morphometric characteristics of the Palar River basin, southern peninsular India; using bAd calculator (bearing azimuth and drainage) and GIS. Geology, Ecology, and Landscapes. 2019;**3**(4):295-307

[30] Singh S. Geomorphology. Allahabad: Prayag Pustak Bhawan; 2000. p. 642

[31] Deen M. Geomorphology and Land Use: A Case Study of Mewat, Thesis Submitted to the Center for the Study of Regional Development. New Delhi: JNU; 1982

[32] Gupta RP. Remote Sensing Geology. Springer; 2018

[33] Strahler AN. Part II. Quantitative geomorphology of drainage basins and channel networks. In: Handbook of Applied Hydrology. New York: McGraw-Hill; 1964. pp. 4-39

[34] Strahler AN. Hypsometric (area-altitude) analysis of erosional topography. Geological Society of America Bulletin. 1952;**63**(11):1117-1142

[35] Talampas WD, Cabahug RR. Catchment characterization to understand flooding in Cagayan De Oro River basin in northern Mindanao, Philippines. Mindanao. Journal of Science and Technology. 2015;**13**

Chapter 8

Marine Applications: The Future of Autonomous Maritime Transportation and Logistics

Nistor Andrei and Cezar Scarlat

Abstract

This chapter presents an in-depth exploration of the maritime industry's transition toward autonomous operations, focusing on the significant advancements in geospatial technologies and artificial intelligence (AI). It examines the evolution of maritime transport, emphasizing the integration of innovative IT solutions, including GIS-based sensors and route optimization techniques. The chapter discusses the emergence and impact of autonomous surface vessels (ASVs), highlighting the challenges they bring in cybersecurity and the necessity for comprehensive standards. Additionally, it explores the potential of a vessel traffic system architecture, with a central node based on an ashore operations center to efficiently manage autonomous maritime logistics. By integrating cutting-edge sensors and AI into ASV systems, the chapter showcases the industry's shift toward a more sustainable, efficient, and technologically advanced future, while addressing the critical issues of safety, security, and environmental impact in this rapidly evolving landscape.

Keywords: maritime transportation and logistics, maritime route optimization, autonomous surface vessels (ASV), ashore operations center (AOC), intelligent sensors, geospatial technologies, artificial intelligence (AI)

1. Introduction

Maritime transportation remains a cornerstone of global trade, facilitating the movement of a vast majority of goods across international borders. Its efficiency and capacity for large-volume transportation make it indispensable in the global economy. Maritime transportation is the backbone of global trade, responsible for carrying approximately 80% of all merchandise by volume [1]. Its critical role in the economy cannot be overstated, as it facilitates the efficient, large-scale movement of goods across the globe. This chapter will examine the significance of this sector and how it continues to adapt and evolve in response to modern challenges and advancements.

The shipping industry has entered a new era characterized by rapid technological advancements and digitalization [2, 3]. These changes are reshaping maritime cargo transportation and logistics, introducing unprecedented levels of efficiency, safety, and environmental sustainability. The trend toward autonomous operations in maritime vehicles, including surface ships and underwater vessels, poses unique

IntechOpen

challenges. These vehicles, unlike their terrestrial counterparts such as cars and trains, have less maneuverability due to their large sizes and masses, along with the complex and ever-changing external disturbances of the sea. This complexity adds significant difficulty in enhancing their autonomy levels, a challenge not as pronounced in ground vehicles [4].

The drive toward autonomy in maritime operations is fueled by the need to enhance safety and efficiency. Offshore operations, often conducted in hazardous environments, pose risks due to variable sea conditions and demanding living conditions. The integration of artificial intelligence systems aboard ships can greatly reduce these risks, replacing much of the offshore work with safer, onshore operations. This transition not only promises improved safety but also offers economic benefits through reduced labor costs. Furthermore, autonomy in maritime vessels is essential in reducing human error, a major factor in maritime accidents.

This chapter aims to investigate the transition from automation to autonomy in maritime transport. It seeks to answer key research questions: how is the maritime industry adapting to technological advancements? What are the stages of this transition, and what implications do they have for future operations? The chapter will provide insights into these questions, offering a comprehensive analysis of the current trends and future prospects in maritime logistics.

2. Current state of maritime automation

The current state of maritime automation is a multifaceted landscape, marked by a dynamic blend of traditional practices and advanced technological innovations. At its core, the industry heavily relies on automated technologies, primarily focusing on navigational aids, cargo handling, and vessel management systems [5, 6]. These innovations have brought about significant improvements in terms of efficiency, safety, and operational consistency.

In the realm of navigational aids, tools like GPS and AIS (automatic identification systems) have become indispensable [7]. They provide vital data that enhance maritime safety and operational efficiency. Vessel management systems, evolving in sophistication, now allow for intricate route planning, fuel management, and comprehensive monitoring of vessel performance [8]. This level of sophistication in vessel management is a leap forward in maritime operations, offering a more nuanced and controlled navigation experience.

Furthermore, automation's impact is profoundly felt in cargo handling processes. The introduction of automated container terminals and robotic handling equipment has revolutionized port operations [9]. These advancements have not only reduced turnaround times but have significantly increased the throughput capacity of ports. This shift toward automation in cargo handling has made ports more efficient and less reliant on intensive human labor.

Despite these advancements, the maritime industry confronts various challenges. The integration of these automated systems into a seamless operational framework remains a complex task. Issues such as cybersecurity vulnerabilities, the need for specialized personnel to manage and oversee these sophisticated systems, and the significant costs associated with implementing such technologies pose considerable challenges.

Yet, these existing automated systems lay a foundational framework for the next evolutionary step in maritime operations: the move toward full autonomy. **Table 1**

Stage	Current state	Interim	Interim	Final state
Port/vessel operation	Manned port/ manned vessel	AOC/manned vessel	Manned port/ ASV	ASV/AOC
Navigation and traffic	Nonautomated traffic	Automated routing, manually navigated by the ship	Autonomous ship in the independent mode	Fully autonomous system
Human involvement	Manual actions mandatory on both sides	Ship-manned work required	Port-manned work required	Human supervision may be emplaced
Key technologies	Basic navigation and communication systems	Advanced routing algorithms, AIS integration	Advanced sensors and autonomous navigation systems for ASV	Comprehensive AI systems and advanced sensors for complete vessel and port autonomy
Challenges	Operational inefficiencies, higher risk of human error	Transitioning to automated systems, training for AOC operations	Integrating ASVs with conventional operations, ensuring safety	Ensuring system reliability, cybersecurity, and seamless integration

Table 1.
The transition from current maritime operations to fully autonomous systems.

offers a structured overview of the transition from current maritime operations to fully autonomous systems, detailing the changes in port/vessel operations, navigation, human involvement, key technologies, and challenges at each stage. This transition is being bolstered by the industry's substantial digital transformation. Shipping companies and ports are increasingly adopting advanced digital technologies, extending beyond navigational and operational applications to encompass the digitalization of entire supply chains. This enhances the overall efficiency, transparency, and traceability of maritime logistics.

Central to this digital transformation are geospatial technologies, especially GIS-based systems [10]. These technologies have become crucial in achieving higher levels of precision and control in maritime operations, enabling advanced route optimization and location tracking. The integration of these systems, both onboard ships and ashore, is a testament to the increasing push toward digitalization in the maritime sector.

The drive toward digitalization is not just an end in itself but a stepping stone toward achieving complete autonomy in maritime transport. This involves the development of AI-controlled systems capable of managing complex navigational and operational tasks. The goal is to enhance safety, reliability, and decision-making efficiency in maritime transport, a domain traditionally reliant on human expertise and judgment.

3. Transitioning toward autonomy: interim states

The maritime sector is actively seeking advanced technologies that can be integrated to lower operational costs. However, the adoption rate of these new technologies varies among ports globally. This variance is primarily due to differences in economic strength, regulatory frameworks, and technological infrastructure among regions [11]. While some areas rapidly embrace these innovations, others proceed cautiously or retain traditional operations due to constraints or strategic choices.

This hybrid state of adoption poses unique challenges for international maritime logistics. Coordination and standardization become more complex as vessels and cargo move between regions with differing levels of technological integration. This scenario necessitates adaptable strategies that can bridge these operational gaps.

The uneven adoption of maritime technologies presents a mixed scenario. On one hand, regions leading in adoption become testbeds for innovation, pushing the boundaries of maritime capabilities. On the other hand, regions that lag behind might face challenges in competitiveness, yet they can benefit from the lessons learned in advanced areas. In **Table 2**, the differences between a port that fully integrates technology and a traditional port are highlighted.

For industry stakeholders, navigating this transitional phase requires a balanced approach. Adopting flexible operational models and engaging in international collaboration for setting standards and best practices can help mitigate the disparities. This approach allows for a smoother transition and paves the way for a more harmonized global maritime future.

As maritime technologies evolve and regions adopt them at different rates, the need for international standardization becomes critical. Standardization ensures compatibility and safety across different systems and regions, facilitating smoother global operations. It is also essential for addressing challenges that arise from the hybrid state of technology adoption in the maritime industry, helping to bridge operational gaps and ensure a coherent approach to maritime autonomy worldwide. Discussing standardization strategies and international collaborative efforts would be a key aspect of navigating the transition toward fully autonomous maritime operations.

Feature	Fully integrated port	Hybrid/traditional port
Vessel operation	ASVs with remote oversight	Manned vessels with crew navigation
Cargo handling	Automated cranes and loading systems	Manual cranes with operator control
Coordination	Centralized through AOC	Decentralized, with on-site management and communication
Traffic management	Managed by advanced vessel traffic systems (VTS) with AI assistance	Conventional VTS with manual monitoring
Dockside operations	Sensor-driven and automated for efficiency and safety	Labor-intensive, with hands-on cargo management
Safety and monitoring	Continuous digital surveillance and predictive analytics	Standard safety protocols with reactive measures
Data handling	Real-time data processing and analytics for operational decisions	Manual data logging and periodic reviews
Human role	Supervisory and exception handling	Direct operation and manual labor
Cost implications	Higher initial investment but reduced operational costs	Lower initial cost but higher long-term operational expenses

Table 2.
Comparison of fully integrated port with a hybrid/traditional one.

However, with digitalization comes a new set of challenges, particularly in the realm of cybersecurity. The maritime industry is increasingly struggling with threats such as data manipulation, spoofing, and cyber attacks. These threats necessitate the implementation of robust security measures to safeguard the integrity of maritime operations.

Looking ahead, the future of vessel traffic systems (VTS) is seen as integral to the successful implementation of maritime autonomy. The development of innovative technologies and methodologies for VTS is essential. These advancements will manage and facilitate the complex traffic of autonomous vessels, ensuring safe and efficient maritime operations. This forward-looking perspective underscores the industry's commitment to not just adapting to technological changes but also shaping them to meet the unique demands of maritime logistics.

4. Fully autonomous vessel traffic system

The integration of AI (particularly deep learning) marks a revolutionary step in maritime operations, enabling vessels to autonomously navigate and operate. AI systems on these vessels can process vast amounts of data from sensors and satellite information to make real-time navigation decisions, adjust routes, and handle complex maritime traffic scenarios. This technology ensures that vessels can operate safely, adhering to maritime laws and regulations without direct human input. Deep learning algorithms continuously improve these processes by learning from past voyages, enhancing efficiency, and safety with each journey.

Incorporating deep learning algorithms, particularly artificial neural networks (ANNs), autonomous vessels achieve remarkable levels of self-navigation and decision-making. These networks, by processing vast quantities of sensory and environmental data, enable vessels to make real-time navigational adjustments, ensuring adherence to maritime safety standards without direct human oversight.

Expanding on this, the interconnected network of neural networks across various vessel systems facilitates a comprehensive and cohesive learning environment. This network synergizes data from multiple sources, enhancing the vessel's ability to respond to complex maritime scenarios. Such an advanced AI integration not only elevates the safety and efficiency of autonomous maritime operations but also symbolizes the forefront of technological innovation in the industry.

In the realm of autonomous maritime operations, advanced sensing and communication play crucial roles. Precise sensors on vessels gather comprehensive environmental and navigational data, facilitating real-time decision-making and course adjustments. These sensors detect everything from oceanic weather patterns to nearby vessel traffic, ensuring safe and efficient navigation.

The integration of advanced sensing in autonomous maritime systems includes both onboard sensors and port-controlled sensors. Onboard sensors provide real-time environmental and navigational data crucial for autonomous decision-making. Simultaneously, port-controlled sensors facilitate the management of incoming and outgoing traffic, ensuring seamless docking and cargo handling. This dual-layered sensor network, coupled with robust communication channels, enables a high degree of synchronization between vessels and port operations, essential for the efficiency and safety of autonomous maritime systems.

For onboard sensors in autonomous maritime systems, the primary requirements are reliability, accuracy, energy efficiency, and cost-effectiveness. The sensor array should include GNSS, AIS, Sonar, EO, IR, and LiDAR for environmental awareness and collision avoidance. Integrating automotive sensors like synthetic aperture radar (SAR), used in automated driving, into maritime systems offers several advantages. SAR's high-resolution imaging capabilities, effective in various visibility conditions, make it ideal for maritime use. This technology can enhance detection and imaging in challenging sea environments, contributing to safer navigation and collision avoidance. Additionally, inertial, velocity, and acceleration measurement sensors are crucial. These sensors must provide accurate, timely data to the navigation AI for tasks like evasive maneuvers and optimal route planning, especially in heavy traffic or docking operations. Ashore sensors, essential for the Ashore Operations Center (AOC), should be equally reliable and accurate. Technologies like SAR are favored for their high-resolution imaging capabilities in various visibility conditions, supporting a network of land, sea, air, and space-based sensors for redundancy and accuracy.

Furthermore, robust communication networks enable seamless data exchange between vessels and port facilities. This connectivity ensures that vessels are continually updated with the latest port conditions and logistics requirements, allowing for optimized docking and cargo handling. This integration of advanced sensing and communication is vital for the smooth operation of autonomous maritime systems, ensuring that vessels not only navigate safely but also synchronize effectively with port operations.

The integration of FPGA (field-programmable gate array) technology can be considered in the design of autonomous maritime systems, especially vessels (due to the compact build-up of FPGA, ensuring space-saving). FPGAs, known for their high-speed processing capabilities and flexibility, can be a key element in managing the complex data processing required for real-time navigation and communication in autonomous vessels. They offer the necessary computational power to handle the vast amount of data from onboard sensors, ensuring swift and accurate decision-making, which is crucial for the safety and efficiency of autonomous maritime operations.

We propose an advanced architecture for artificial neural networks (ANNs) in autonomous maritime systems, utilizing a "many-to-many" recurrent neural network. This network integrates elements such as ship AI (covering navigation, operation, and sensor control) and the ashore operations center AI (managing traffic, ship control, sensor control, and providing a common operational picture). The recurrent nature of this network facilitates feedback between hidden layers and across all neurons. For hardware implementation, field-programmable gate array (FPGA) boards are suggested due to their robust parallel computation capabilities and low latency, crucial for handling time-sensitive sensor signals. This system can also be supported by CPUs or GPUs, maintaining the specified properties. The overall structure is conceptualized as a network of networks, or a system of systems, ensuring comprehensive and integrated functionality.

Automated navigation systems in autonomous vessels are integral for efficient maritime operation. These systems utilize a combination of AI, advanced sensors, and satellite data to autonomously chart and adjust courses. They continuously assess environmental factors like weather and sea conditions, ensuring safe and optimized navigation. This autonomous capability allows vessels to efficiently reach their destinations, effectively reducing human error and enhancing maritime safety.

Energy management in autonomous maritime systems plays a pivotal role in promoting sustainability. These systems optimize fuel use and increase energy efficiency by employing advanced algorithms. Such optimization leads to reduced greenhouse gas emissions and a lower ecological footprint, aligning maritime operations with

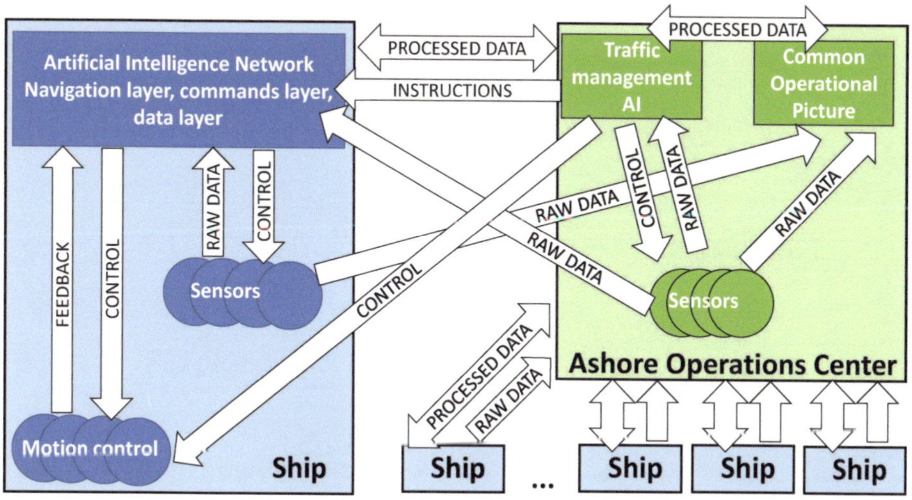

Figure 1.
Diagram of the proposed autonomous traffic management system.

environmental conservation goals. This approach not only improves operational efficiency but also contributes significantly to greener and more sustainable maritime operations (**Figure 1**).

Predictive maintenance in autonomous maritime systems is a forward-thinking approach that enhances vessel reliability and longevity. These systems use onboard sensors and data analytics to monitor vessel conditions, identifying potential maintenance issues before they escalate. This proactive maintenance strategy minimizes unexpected downtime, ensuring consistent and efficient vessel operations. By scheduling repairs and maintenance in advance, the lifespan of key components is extended, and operational efficiency is maintained.

Autonomous cargo handling in autonomous maritime systems revolutionizes the loading and unloading process. These systems utilize advanced automation technology for cargo handling, ensuring both speed and precision [12]. The use of automated cranes and conveyor systems, guided by sensors and AI algorithms, allows for accurate placement and movement of cargo. This automation not only streamlines the process, reducing the time spent in port, but also minimizes the potential for human error, enhancing the overall operational efficiency in maritime logistics.

Regulatory compliance in autonomous maritime systems involves ensuring that these vessels adhere strictly to international maritime laws and standards. This involves implementing protocols within the vessel's control systems that continuously monitor and adjust operations to stay within legal and safety guidelines. The integration of these protocols is crucial for maintaining global maritime safety standards and ensuring that autonomous vessels operate within the legal framework of international maritime law, ensuring safe and compliant maritime operations.

5. Challenges and considerations

In the rapidly evolving landscape of maritime technology, cybersecurity has emerged as a major concern for autonomous surface vessels and port operation centers. The transition to autonomy in maritime operations presents unique

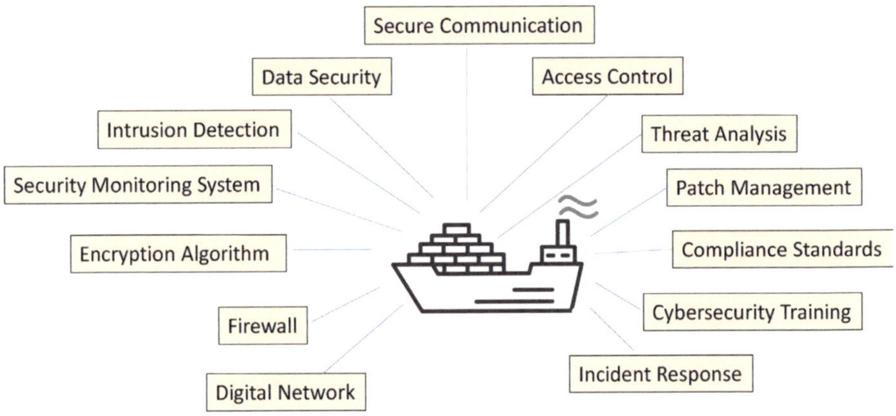

Figure 2.
Diagram that illustrates cybersecurity measures for autonomous vessel traffic systems.

cybersecurity challenges, given the reliance on complex communication and data systems. These systems, central to the operational efficiency and safety of autonomous vessels, are susceptible to a range of cyber threats [13]. This necessitates a robust and comprehensive approach to cybersecurity, ensuring the integrity, confidentiality, and availability of vital information. Protecting these vessels from cyber threats is not just about safeguarding assets; it is integral to maintaining maritime safety, environmental protection, and the security of global supply chains. **Figure 2** presents a range of cybersecurity measures that can be adopted in autonomous maritime operations.

In this context, enhanced encryption and data security stand as crucial tools against cyber threats. As vessels and operation centers increasingly rely on digital communication, the risk of data breaches and unauthorized access escalates. Advanced encryption methods are essential to secure these communications, ensuring that sensitive data transmitted between vessels and shore-based centers remain confidential and tamper-proof. Encryption technologies are suitable for maritime applications, and they have a significant role in protecting against cyber espionage, data manipulation, and other malicious activities that could compromise the safety and efficiency of autonomous maritime operations.

Continuing from the necessity of advanced encryption methods in securing communications, it is imperative to integrate these technologies within the complex network of data exchange between AOC and ASV or ASV and other ASV. This integration is critical for ensuring the accuracy and timeliness of data crucial for the operational efficiency of port operations. Real-time monitoring systems and advanced encryption protocols can work in tandem to safeguard this data exchange from cyber-attacks. Such measures are indispensable for enabling the ANNs to generate reliable commands and validate routes effectively, thereby ensuring seamless maritime operations with minimal risks.

The integration of artificial intelligence and deep learning for threat detection marks a significant advancement in the cybersecurity of autonomous maritime operations. These technologies offer the ability to analyze vast amount of data rapidly, identifying and responding to potential cyber threats more efficiently than traditional methods. By leveraging AI, autonomous vessels and operation centers can detect anomalies and potential security breaches in real time, enabling proactive measures to mitigate risks.

Establishing robust network security protocols is vital for the integrity of autonomous maritime operations. The interconnected nature of these systems makes them vulnerable to network-based cyber-attacks, such as intrusions or distributed denial of service (DDoS) attacks. By implementing stringent network security measures, autonomous vessels and operation centers can safeguard their critical infrastructure. It is crucial to include in the design of the system essential components of these protocols, such as firewalls, intrusion detection systems, and secure network architectures, which are key in preventing unauthorized access and ensuring the resilience of maritime communication networks against cyber threats.

Continuous monitoring systems are a cornerstone in the cybersecurity strategy for autonomous surface vessels and operation centers. These systems play a critical role in the early detection and response to cyber threats, ensuring the ongoing integrity and security of maritime operations. Continuous monitoring involves the constant scrutiny of network traffic, system logs, and operational patterns to identify unusual activities that may signify a cybersecurity breach. These monitoring systems can provide real-time alerts, enabling swift action to counteract potential threats and maintain the secure operation of autonomous maritime systems.

Another way to maintain the cybersecurity of autonomous maritime systems is through regular updates of software and firmware are imperative for. These updates often include patches for newly discovered vulnerabilities, which, if left unaddressed, could be exploited by cyber-attackers. In the context of autonomous vessels and operation centers, staying in line with the latest software updates ensures that the systems are fortified against emerging threats. By developing strategies for implementing a rigorous update protocol, timely updates are ensured in order to safeguard the complex and interconnected systems that are vital for the safe and efficient operation of autonomous maritime vessels.

Cybersecurity training for maritime personnel is also a critical element in safeguarding autonomous maritime systems. The human factor plays a significant role in maintaining cybersecurity; thus, equipping staff and crew with the necessary knowledge and skills is paramount. This training should cover best practices in cybersecurity, threat awareness, and protocols for responding to cyber incidents. Such educational initiatives help in building a culture of cybersecurity awareness within maritime operations, significantly reducing the risk of breaches caused by human error.

Implementing a multilayered defense strategy is essential in the cybersecurity of autonomous maritime systems. This approach combines various security measures at different operational levels to create a comprehensive defense against cyber threats. It includes physical security of hardware, network defenses, and application-level safeguards. Each layer serves as a barrier, ensuring that even if one level is compromised, additional layers of security protect the system. By implementing a diversified defense strategy, a resilient and robust cybersecurity framework is created, crucial for the complex and interconnected nature of autonomous maritime operations.

Developing effective incident response plans is a critical aspect of cybersecurity in autonomous maritime operations. These plans outline procedures to be followed in the event of a cyber-attack, ensuring a coordinated and efficient response. An effective plan includes identifying potential threats, establishing communication protocols, and defining roles and responsibilities during a security incident.

Collaboration within the maritime industry and the adoption of blockchain technology are forward-looking approaches to enhancing cybersecurity for autonomous vessels and operation centers [14]. Collaborating with industry peers and government entities allows for the sharing of best practices, threat intelligence, and joint development of security standards. Blockchain technology, with its inherent features of decentralization, transparency, and tamper-resistance offers a novel approach to secure record-keeping and communication in maritime operations [3]. The adoption of collaborative efforts and advanced technologies like blockchain represents a paradigm shift in maritime cybersecurity. By engaging in cross-sector collaboration, stakeholders in the maritime industry can leverage collective expertise and resources to develop more robust cybersecurity measures. Blockchain technology, with its capacity for secure and transparent data management, offers a revolutionary approach to safeguarding communication and operational data in autonomous maritime systems. This collaborative and technological synergy paves the way for a new era in maritime cybersecurity, characterized by enhanced resilience and adaptability to evolving cyber threats.

In the autonomous maritime system with the architecture proposed in Section 4, the incorporation of advanced cybersecurity measures is fundamental. This includes integrating enhanced encryption for secure communication between ASVs, AOCs, and other vessels. Leveraging AI and deep learning enables proactive threat detection and response, ensuring real-time protection against cyber-attacks. Robust network security protocols, continuous monitoring, and regular updates of software and firmware are crucial to safeguard the system's integrity. Additionally, comprehensive cybersecurity training for maritime personnel and a multilayered defense strategy fortify the system against evolving cyber threats. These measures are integral to ensuring the resilience and seamless operation of the autonomous maritime network.

6. The future of maritime industry

In the future of autonomous maritime transportation, the integration of advanced artificial intelligence and machine learning (ML) is pivotal. These technologies will revolutionize navigation systems, making them more precise and efficient. AI algorithms could predict and analyze various maritime conditions, facilitating real-time decision-making for safer and more efficient voyages. Additionally, AI and ML can significantly enhance predictive maintenance, reducing downtime and increasing the longevity of vessel components. This integration symbolizes a shift toward a smarter, more autonomous maritime industry, where decision-making is data-driven and adaptive to changing maritime environments.

The deployment of the Internet of Things (IoT) and sensor networks in autonomous maritime transportation represents a transformative step. These technologies will enable a more interconnected and responsive maritime environment, where real-time data from various sensors on the vessel can be continuously monitored and analyzed. This network of sensors and IoT devices will provide critical information on vessel health, environmental conditions, and operational status, thereby enhancing situational awareness and predictive analytics. The integration of such a connected ecosystem is fundamental to the evolution of smarter, more efficient maritime operations, where data-driven insights lead to optimized performance and safety.

Sustainable energy solutions will be a key focus for the autonomous maritime industry. Innovations in renewable energy sources such as solar, wind, and hybrid propulsion systems will be increasingly adopted to power autonomous surface vessels (ASVs). This shift not only addresses environmental concerns but also enhances the energy efficiency of maritime operations. The use of these sustainable energy technologies will be crucial in reducing the carbon footprint of shipping, aligning the maritime industry with global environmental sustainability goals. The move toward greener energy sources is a significant step in the evolution of the maritime sector, contributing to a cleaner and more sustainable maritime environment.

The enhancement of communication systems will play a crucial role in the advancement of autonomous maritime operations. Future communication technologies need to be more robust and capable of handling the complex data exchange between ASVs and shore-based operation centers. This includes high-bandwidth, low-latency communication channels to facilitate real-time data transfer and decision-making processes. Improved communication systems will not only increase the operational efficiency of ASVs but also ensure a higher level of safety by enabling better coordination and response to maritime incidents.

Developing effective safety and risk management protocols will be essential in the autonomous maritime industry. As vessels operate with increased autonomy, there's a need for robust safety standards and risk assessment models specifically designed for these advanced systems. These protocols will encompass not just the technological aspects but also the human factors in automated environments. The aim is to establish a comprehensive safety framework that anticipates potential risks, mitigates hazards, and ensures a high level of safety in autonomous maritime operations.

As autonomous maritime operations evolve, the role of human operators will shift toward a more collaborative model with autonomous systems. This human-machine collaboration will focus on leveraging the strengths of both – the advanced analytical capabilities of machines and the intuitive decision-making skills of humans. This synergy is essential for optimizing performance and ensuring safety in complex maritime environments. The future will likely see humans in supervisory and decision-support roles, ensuring that autonomous systems function within safe and efficient parameters. This collaborative approach is key to the successful integration of autonomy in maritime operations.

The development and implementation of an international regulatory framework will be crucial for the future of autonomous maritime operations. This framework will need to address the unique challenges posed by autonomous vessels, ensuring standardized safety, communication, and operational protocols. Establishing such regulations will require collaboration among maritime nations, industry stakeholders, and international maritime organizations. The goal is to create a consistent and comprehensive set of rules and standards that govern the deployment and operation of autonomous vessels, ensuring safety and efficiency across global maritime trade routes.

Enhancing cybersecurity measures is a key aspect of the future of autonomous maritime operations. As vessels become more interconnected and reliant on digital systems, the risk of cyber threats increases. It is crucial to continuously develop and implement advanced cybersecurity protocols to protect against these threats. This will involve not only deploying sophisticated technology to safeguard systems and data but also ensuring regular updates and monitoring to stay ahead of potential cyber vulnerabilities. The focus will be on creating a secure digital environment that supports the safe and efficient operation of autonomous vessels.

The future of maritime traffic management systems will involve significant innovations, particularly in the integration of autonomous technologies. These systems will become more sophisticated and capable of managing complex operations involving both manned and unmanned vessels. The focus will be on enhancing the efficiency and safety of maritime traffic through advanced algorithms and real-time data processing. This evolution is crucial for handling the increasing volume of maritime traffic and the complexities introduced by autonomous vessels, ensuring smooth and safe navigation in busy sea routes.

Continued investment in research and development (R&D) will be pivotal for driving advancements in the maritime industry, especially in the realm of autonomous operations. This investment is not only crucial for technological innovation but also for addressing emerging challenges and adapting to evolving industry standards. The focus will be on developing more efficient, safe, and environmental friendly autonomous systems. R&D will play a key role in ensuring that the maritime industry remains at the forefront of innovation, constantly pushing the boundaries of what is possible in the autonomous maritime technology.

7. Conclusions

The integration of technologies like AI, IoT, and advanced sensors in the maritime industry marks a revolutionary shift. These innovations not only enhance navigational accuracy and operational efficiency but also open new avenues for autonomous vessel management. By analyzing vast amounts of data, AI and machine learning enable predictive maintenance and real-time decision-making, making maritime operations safer and more reliable.

AI's role in transforming maritime operations cannot be overstated. It brings a new level of intelligence to autonomous vessels, from optimizing routes to ensuring safety. The ability of AI to process and react to dynamic maritime environments ensures that autonomous vessels can operate effectively in varying conditions. This adaptability is key to the success of autonomous maritime systems in the future.

The implementation of IoT technologies in maritime operations heralds a new era of connectivity. Sensor networks on vessels provide critical data on vessel performance and environmental conditions, facilitating enhanced monitoring and control. This interconnectedness ensures seamless communication between vessels and shore-based operations, laying the groundwork for a more integrated and responsive maritime industry.

In the era of autonomous maritime operations, the emphasis on cybersecurity cannot be overstated. As vessels become more connected and reliant on digital technologies, they are increasingly susceptible to cyber threats. Implementing advanced cybersecurity measures is crucial to protect these systems from unauthorized access and cyberattacks. This not only safeguards the vessels and their cargo but also ensures the safety of maritime personnel and the marine environment.

The need for proactive cybersecurity strategies in autonomous maritime operations is imperative. Continuous monitoring, regular software updates, and employee training in cybersecurity best practices are essential components of a robust cybersecurity framework. These measures, combined with state-of-the-art encryption and intrusion detection systems, form the backbone of a secure autonomous maritime infrastructure, ensuring the resilience of these systems against evolving cyber threats.

As autonomous maritime technologies advance, the role of human operators shifts toward more strategic oversight and decision-making. The human element remains crucial for interpreting data and handling complex situations where machine algorithms might falter. This collaboration between human expertise and machine efficiency ensures a balanced approach to maritime operations, enhancing both safety and operational effectiveness. The integration of human-machine collaboration underscores the need for stringent safety protocols. These protocols not only ensure the smooth functioning of autonomous systems but also safeguard human operators. Regular training, emergency preparedness, and adherence to safety standards are vital in managing the unique risks associated with autonomous maritime technologies. This human-centric approach to safety in an increasingly automated industry is key to maintaining a secure maritime environment.

With rapid advancements in autonomous maritime technologies, there is a pressing need for international regulatory bodies to update and create new standards. These regulations must encompass the unique aspects of autonomous operations, addressing safety, communication, and environmental concerns. Keeping pace with technological innovation is crucial for maintaining a safe and efficient global maritime framework. The evolution of international regulations requires collaborative efforts among maritime nations and organizations. Standardizing rules for autonomous maritime operations ensure consistency in safety and operational procedures worldwide. This collective approach is vital for fostering an environment where autonomous maritime technologies can be seamlessly integrated and universally adopted.

The autonomous maritime sector must continually embrace innovation to stay ahead of technological advancements and emerging challenges. This ongoing innovation is essential for improving operational efficiency, enhancing safety standards, and reducing environmental impact. It requires a commitment to research and development, fostering a culture that continuously seeks to improve and evolve. The future of the autonomous maritime industry relies heavily on collaboration between industry players, academia, and regulatory bodies. Joint efforts in research, sharing of best practices, and development of unified standards are crucial for harnessing the full potential of autonomous technologies. Looking ahead, the maritime sector must proactively address potential challenges such as cybersecurity threats, regulatory compliance, and ethical considerations. Anticipating these challenges and developing strategies to mitigate them ensures the industry is well-prepared to adapt and thrive in an increasingly automated future.

Conflict of interest

The authors declare no conflict of interest.

Author details

Nistor Andrei* and Cezar Scarlat
Doctoral School of Entrepreneurship, Business Engineering and Management, National University for Science and Technology POLITEHNICA Bucharest, Romania

*Address all correspondence to: nandrei.upb@gmail.com

References

[1] MSC. Report of the Maritime Safety Committee on its One Hundredth Session [Internet]. Geneva: International Maritime Organization; 2022. Available from: https://www.mpa.gov.sg/docs/mpalibraries/mpa-documents-files/shipping-division/msc-reports/msc-100-20-add-1---report-of-the-maritime-safety-committee-on-its-one-hundredth-session-(secretariat).pdf [Accessed: April 15, 2023]

[2] Aslam S, Michaelides MP, Herodotou H. Internet of ships: A survey on architectures, emerging applications, and challenges. IEEE Internet of Things Journal. 2020;7(10):9714-9727

[3] Kuo SY, Huang XR, Chen LB. Smart ports: Sustainable smart business port operation schemes based on the artificial intelligence of things and blockchain technologies. IEEE Potentials. 2022;**41**(6):32-37

[4] Aiello G, Giallanza A, Mascarella G. Towards shipping 4.0. A preliminary gap analysis. Procedia Manufacturing. 2020;**42**:24-29

[5] Weintrit A. Clarification, systematization and general classification of electronic chart systems and electronic navigational charts used in marine navigation. Part 1 - electronic chart systems. TransNav. 2018;**12**(3):471-482

[6] Porretta M, Jiménez-Baños D, Crisci M, Fiumara A, Hein G. GNSS evolutions for maritime: An incremental approach. InsideGNSS. 2016;**10**(May/June):54-62

[7] AIS transponders [Internet]. Available from: https://www.imo.org/en/OurWork/Safety/Pages/AIS.aspx [Accessed: January 8, 2024]

[8] Horowitz J. Global Shipping Needs Geospatial Technology. xyHt Digital Magazine [Internet]; 2020; Available from: https://www.xyht.com/gnsslocation-tech/global-shipping-needs-geospatial-technology/

[9] Hu ZH, Sheu JB, Luo JX. Sequencing twin automated stacking cranes in a block at automated container terminal. Transportation Research Part C: Emerging Technologies. 2016;**69**:208-227

[10] Baylon A, Santos EM. Introducing GIS to TransNav and its extensive maritime application: An innovative tool for intelligent decision making? TransNav. 2013;7(4):557-566

[11] Jia X, Cui Y. Examining interrelationships of barriers in the evolution of maritime port smartification from a systematic perspective. Transport Policy. 2021;**114**:49-58

[12] Merz M, Grøtli EI, Mørkrid OE, Tangstad E, Fossøy S, Nordahl H. A gap analysis for automated cargo handling operations with geared vessels frequenting small sized ports. Maritime Transport Research. 2023;**5**:100098

[13] Cho S, Orye E, Visky G, Prates V. Cybersecurity Considerations in Autonomous Ships. Vol. 1(1). Available from: https://ccdcoe.org/library/publications/cybersecurity-considerations-in-autonomous-ships/. CCDCOE [Internet]; 2022 [Accessed: December 25, 2023]

[14] Dong C, Huang Q, Pan Y, Ng CT, Liu R. Logistics outsourcing: Effects of greenwashing and blockchain technology. In: Transportation Research Part E-Logistics and Transportation Review. Vol. 170. Elsevier; 2023

Chapter 9

Entropy of GPS-measured Earth Tremor

Alexey Lyubushin

Abstract

Modern GPS networks make it possible to study the tremors of the earth's surface from the point of view of identifying anomalous areas. The use of the entropy of the distribution of wavelet coefficients provides a tool for highlighting the hidden and non-obvious properties of the earth's surface tremors. The principal component method makes it possible to identify the most important general trends in the behavior of informative tremor statistics and determine areas of anomalous behavior. The application of these methods to the analysis of GPS data in California is presented. Particular attention is paid to time intervals and areas (clusters) with extreme entropy values. Periodicities in the occurrence of strong jumps in the average entropy of the entire region have been discovered, of which the period of 95 days is dominant. The trend of migration of areas of maximum entropy from the South to the North has been identified. As a result of the analysis, it was found that the area of minimum entropy values gravitates toward the San Andreas fault, and the vicinity of San Francisco has the selected properties of maximum information content and attracts low entropy trajectories.

Keywords: GPS, entropy, wavelets, principal components, correlations, California

1. Introduction

The fine structure of the time series of displacements of the earth's surface measured using GPS is the subject of research by a large number of specialists. The maximum likelihood approach for estimating parameters of GPS time series models was used in [1–4]. In works [5–8], the maximum likelihood method was used to estimate the parameters of the shapes of power spectra and noise amplitude in various regions of the world. The noise structure and error estimates are considered in [7, 9, 10]. Errors in estimating the rate of displacement depending on the shape of the spectrum and the intensity of the noise component of the time series were analyzed in [11, 12]. Parametric models of GPS time series for the analysis of tectonically active areas, including the analysis of phase correlations, were used in [13, 14].

The structure of the high-frequency noise and low-frequency seasonal components of GPS time series in connection with the task of estimating the rates of displacement of tectonic plates was studied in [15–18]. Seasonal effects in GPS signal series associated with hydrological load and tectonic movements of the earth's crust were studied in [19–26] using various methods, including a multivariate statistical approach. The influence of ionospheric and tropospheric delays on the sensitivity of GPS solutions was considered in [27]. Hypotheses about the occurrence of "anomalous" harmonics in GPS time series are considered in [28].

Multivariate statistics methods (principal components, empirical orthogonal functions, and singular spectrum estimation) in [29, 30] were used to identify common spatial and temporal components of GPS time series. A joint analysis of the noise component of time series and accelerometer readings was carried out in [31]. Non-stationary effects were studied in [32–34] to assess the mutual movements of blocks of the earth's crust and the stability of station positioning.

The coherence of ground tremor was analyzed in [35, 36]. In [37, 38], coherence analysis of GPS time series was used to assess seismic hazards in Japan and California.

This chapter presents the further development of methods for analyzing ground tremors proposed in [39, 40] and analyzes the structure of high-frequency noise in time series of a network of GPS stations in the Western United States during the time period 2009–2022. The Western USA (mainly California) is one of the most tectonically active regions. A dense network of GPS stations makes it possible to study in detail the characteristics of vibrations of the earth's surface and try to compare them with known regions of seismic activity. Previously, in [41], a study of the properties of seismic noise in Southern California was carried out in comparison with the seismic regime.

In this work, to analyze daily GPS time series, we use estimates of the entropy of the distribution of the squared wavelet coefficients of the expansion into orthogonal wavelets in a sliding time window. Since, due to the orthogonality of the wavelet transform, the squared modulus of each wavelet coefficient is equal to a certain "atom" of oscillation energy, the entropy of the energy distribution over the complete set of independent time-frequency intervals is actually studied.

The term tremor has many different meanings in geophysics. In seismology, tremor refers to slow movements of the earth's crust as a result of the so-called quiet earthquakes. In volcanology, tremor is a shaking of the earth's surface in the vicinity of volcanoes, usually associated with eruptions. In this chapter, tremor is the increments of daily GPS time series. The transition to increments is a well-known operation in time series analysis, which allows one to reduce the influence of strong low-frequency components and increase the stationarity of the analyzed signals.

2. Data used

Data on earth surface displacements measured by GPS were taken from the Nevada Geodetic Laboratory website at: http://geodesy.unr.edu/gps_timeseries/tenv3/IGS14/ [42].

Positions of 699 GPS stations for a rectangular domain $32.5^\circ \leq Lat \leq 51^\circ$, $-128.5^\circ \leq Lon \leq -115^\circ$ which have daily time series of 5113 samples ranging from the beginning of 2009 to the end of 2022 (14 years), having a total number of gaps of less than 365 samples, and the longest gap of less than 182 samples, are presented at the left panel of **Figure 1**. Vertical components of earth displacement are investigated. Gaps in the GPS time series are filled using information from the left-hand and right-hand vicinities of the gap of the same length as the length of the gap [35].

The set of stations was previously divided into clusters. To divide the network of stations into clusters, 18 reference points were selected (**Figure 1**). The number 18 was chosen as the optimal number of station clusters, which split their "cloud" using the k-means method. Let us partition the set of station position vectors $\vec{\zeta}$ into a given

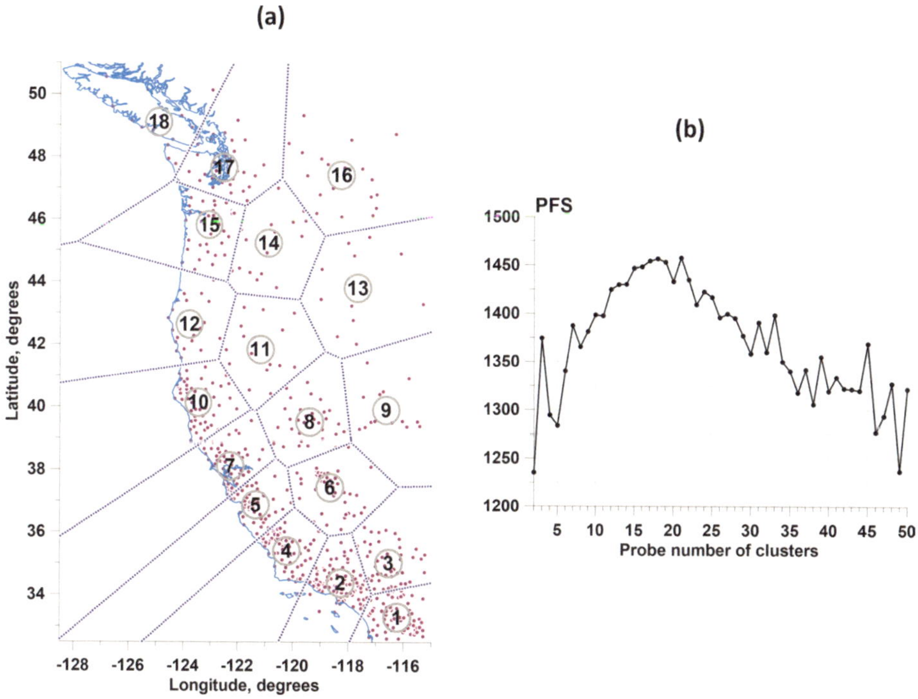

Figure 1.
(a) It presents the positions of 699 GPS stations and their partition into 18 clusters. The numbered circles represent the centers of gravity of the clusters, the blue lines are the boundaries between Voronoi cells. (b) It shows a plot of pseudo-F statistics, which made it possible to select 18 as the number of clusters.

trial number q of clusters using the popular k-means clustering method [43]. Denote by $C_r, r = 1, \dots, q$ clusters, let $\vec{z}_r = \sum_{\vec{\zeta} \in C_r} \vec{\zeta}/n_r$ be the center of mass vector of the cluster C_r, n_r be the number of vectors in the cluster C_r, $\sum_{r=1}^{q} n_r = N$. A vector $\vec{\zeta} \in C_r$ if the distance $|\vec{\zeta} - \vec{z}_r|$ is minimal among the positions of all cluster centers. The k-means method minimizes the sum of distances:

$$G\left(\vec{z}_1, \dots, \vec{z}_q\right) = \sum_{r=1}^{q} \sum_{\vec{\zeta} \in C_r} \left|\vec{\zeta} - \vec{z}_r\right|^2 \rightarrow \min_{\vec{z}_1, \dots, \vec{z}_q} \tag{1}$$

with respect to the position of cluster centers $\vec{z}_r$. Let $\Phi(q) = \min_{\vec{z}_1, \dots, \vec{z}_q} G\left(\vec{z}_1, \dots, \vec{z}_q\right)$. We used a trial number of clusters in the range $2 \le q \le 50$. The problem of choosing the best number of clusters q^* was solved using the pseudo-F-statistic maximum criterion [44]

$$PFS(q) = \sigma_1^2(q)/\sigma_0^2(q) \rightarrow \max_{2 \le q \le 50} \tag{2}$$

where

$$\sigma_0^2(q) = \frac{\Phi(q)}{N - q}, \quad \sigma_1^2(q) = \sum_{r=1}^{q} \frac{n_r}{N} \cdot \left|\vec{z}_r - \vec{z}_0\right|^2, \quad \vec{z}_0 = \frac{1}{N} \sum_{1}^{N} \vec{\zeta} \tag{3}$$

Clust#	1	2	3	4	5	6	7	8	9	10	11	12	13	14	15	16	17	18
N_{sta}	65	97	46	66	50	52	57	32	14	50	15	20	11	17	39	19	35	14

Table 1.
Number of N_{sta} stations in each Clust# cluster.

The right-hand graph in **Figure 1** presents the values of the pseudo-F-statistic as a function of the trial number of clusters. The number 18 on the graph of the pseudo-F-statistics is the breakpoint of the dependence on the trial number of clusters and realizes the maximum for the number of clusters from 2 to 50. On the graph of the pseudo-F statistics, there are two local maxima with close values for the number of clusters 18 and 21. Of these two values, the number 18 was chosen as the smaller one. In **Figure 1**, the division of a set of stations into 18 clusters is shown along with Voronoi cells, which clearly indicate that the stations belong to one or another cluster.

Clusters of stations are ordered by increasing latitude of the positions of their centers of gravity. **Table 1** shows for each cluster (first line) the number of stations in the cluster (second line).

3. Principal components of increments in a sliding time window

Since the goal is to study the tremor of the earth's surface, that is, the high-frequency part of the earth's surface displacements, the analysis was carried out for increments of time series. The transition to increments reduces the dominant influence of low frequencies in the daily GPS time series and ensures stationarity of time series fragments within the 365-day time windows that are used further.

The division of stations into 18 clusters is used for subsequent application of the principal component method [45]. For each cluster of stations, the first principal component of the time series of increments of vertical displacements of the earth's surface was calculated in a sliding adaptation time window of 365 days in length.

Let us consider several time series $P(t) = (P_1(t), \dots, P_m(t))^T$ of common dimensionality m. In our case $m = 18$. It is necessary to estimate the first principal component in a sliding time window of long L samples. To do this, consider samples with time indexes t under the condition $s - L + 1 \leq t \leq s$, s - the right end of the time window. The correlation matrix $\Phi(s)$ of the size $m \times m$ is calculated using the formulas:

$$\Phi(s) = \left(\varphi^{(s)}_{ab}\right), \quad \varphi^{(s)}_{ab} = \sum_{t=s-L+1}^{s} q^{(s)}_{a}(t) q^{(s)}_{b}(t)/L, \quad a, b = 1, \dots, m \tag{4}$$

where

$$q^{(s)}_{a}(t) = \left(P_a(t) - \overline{P}^{(s)}_{a}\right)/\sigma^{(s)}_{a}, \quad \overline{P}^{(s)}_{a} = \sum_{t=s-L+1}^{s} P_a(t)/L, \quad \left(\sigma^{(s)}_{a}\right)^2 = \\ = \sum_{t=s-L+1}^{s} \left(P_a(t) - \overline{P}^{(s)}_{a}\right)^2/(L-1), \quad a = 1, \dots, m \tag{5}$$

Let $\theta^{(s)} = \left(\theta^{(s)}_{1}, \dots, \theta^{(s)}_{m}\right)^T$ be the eigenvector of the matrix $\Phi(s)$ with the maximum eigenvalue. Let us calculate the value:

$$\psi^{(s)}(t) = \sum_{\alpha=1}^{m} \theta^{(s)}_{a} \cdot q^{(s)}_{a}(t) \tag{6}$$

and define the scalar time series $\psi(t)$ of the adaptive first principal component in a sliding time window in accordance with the formula:

$$\psi(t) = \begin{cases} \psi^{(L-1)}(t), 0 \leq t \leq (L-1) \\ \psi^{(t)}(t), t \geq L \end{cases} \tag{7}$$

Eq. (4)–(7) are applied independently in each time window. According to them, in the first time window, a series $\psi(t)$ consisting of L values, the values of the principal component are calculated according to Eqs. (6)–(7). In all subsequent time windows, the time series values $\psi(t)$ correspond to the single estimates at the far right ends of the respective windows. Thus, outside the first time window, the values of the series $\psi(t)$ depend only on past values $P(t)$. The calculations of the principal components were carried out in a sliding time window of 365 days with a step of 3 days. The length of the adaptation time window of 365 days is equal to 1 year and was chosen for the reason that it coincides with the natural period of variations in GPS time series. After calculating the first principal components, information is compressed and the dimension of the analyzed data is reduced from 699 to 18. **Figure 2** shows the graphs of the first principal components for 4 clusters as an example.

4. Entropy of the first principal components

The minimum normalized wavelet-based entropy for the sample $x(t),\ \ t = 1, .., N$ is determined by the formula:

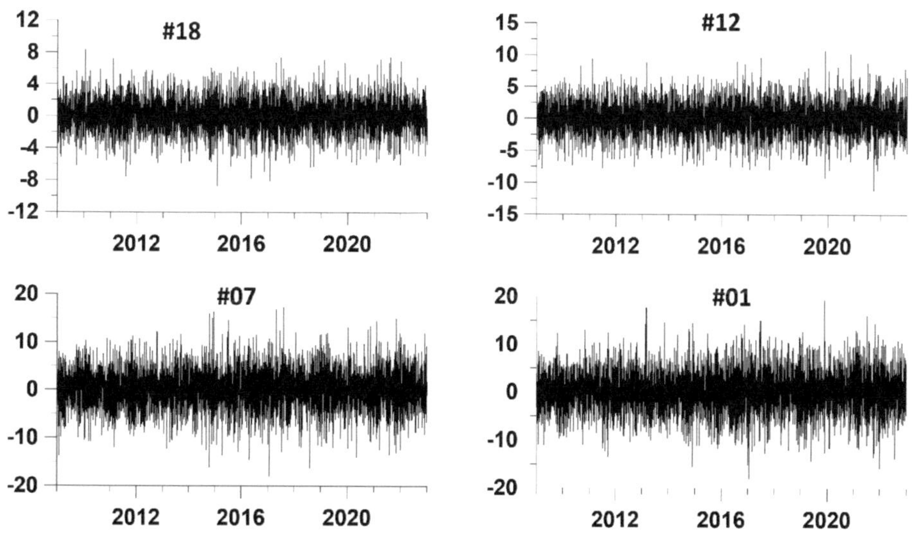

Figure 2.
Graphs of first principal components within moving time window of the length 365 days for 4 clusters.

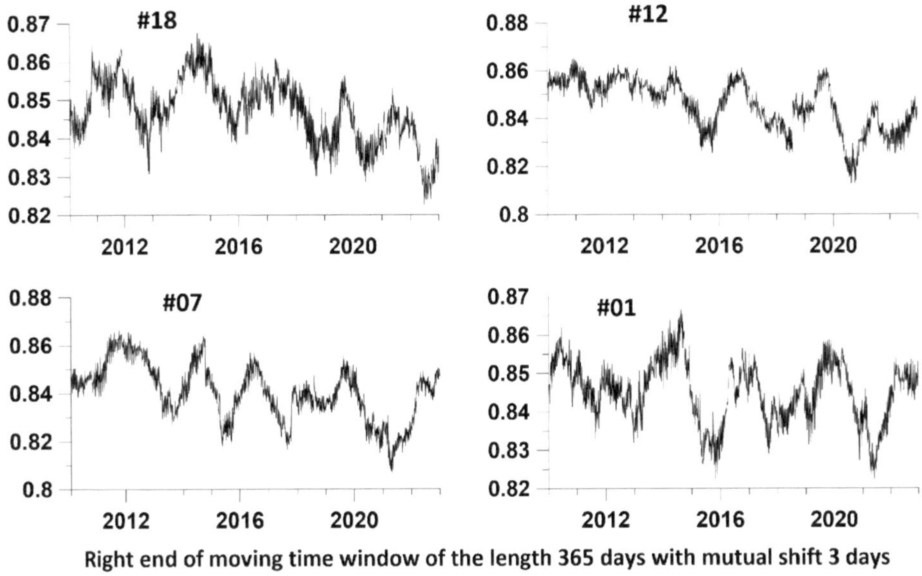

Figure 3.
Graphs of minimum normalized wavelet-based entropy for first principal components within moving time window of the length 365 days with mutual shift 3 days for 4 clusters.

$$En = -\sum_{k=1}^{N} p_k \cdot \log(p_k) / \log(N), \quad p_k = c_k^2 / \sum_{j=1}^{N} c_j^2 \tag{8}$$

Here c_k denoted the coefficients of the orthogonal wavelet decomposition. Daubechies bases were used with the number of vanishing moments from 1 to 10 [46]. We chose such a basis from this set for which the value Eq. (8) is minimal for a given time window and for the considered component of the displacement of the earth's surface. By Eq. (8) $0 \leq En \leq 1$. We will consider estimates of the entropy of principal components of GPS time series increments in a sliding time window 365 days long with a shift of 3 days.

Entropy Eq. (8) has been repeatedly used in [41, 47] to analyze the variability of seismic noise in different regions of the Earth and at the global level as an independent tool. By its design, entropy Eq. (8) is also multiscale, like the entropy proposed in [48] for studying the properties of random signals. In [49], the author used the non-extensive Tsallis entropy to analyze seismic noise. The "natural time" approach to the analysis of random data uses a related definition of entropy in [50].

Figure 3 shows graphs of entropy values for the first principal components of the increments of the vertical components of the displacement of the earth's surface, estimated in a sliding time window of 365 days in length with a displacement of 3 days for 4 clusters of stations, the same as those presented in the previous **Figure 2**.

5. Mean entropy and its wavelet transform maximum modulus stepwise approximation

Entropy is a measure of the complexity of the behavior of a random signal. A simple signal which is close by its properties to white noise has maximum entropy. On the contrary, the entropy of a signal, saturated with non-stationary components and sharp changes in its amplitude, reaches minimum values. Therefore, of particular interest are those time fragments and those areas of the region under study where entropy reaches extreme values, minimums or maximums. Next, to isolate such time fragments, a smoothing method is used, the purpose of which is to approximate the signal with a piecewise step function. The scale-dependent piecewise step approximation makes it possible to formally determine the moments in time of a significant change in the current average signal values. These moments in time can further be considered as a kind of event in the dynamics of a random time series. To construct such an approximation, continuous wavelet transforms with a kernel in the form of derivatives of a Gaussian function are used, which makes it possible to formalize the selection of time points at which a scale-dependent significant change in the average value of the noisy signal occurs [46, 51].

Let $Y(t)$ be an arbitrary signal and consider the corresponding smoothed signal obtained by using a scale-dependent kernel:

$$c_0(t,a) = \int_{-\infty}^{+\infty} Y(t+av)\cdot\psi_0(v)\,dv \Big/ \int_{-\infty}^{+\infty} \psi_0(v)\,dv \tag{9}$$

where $a > 0$ is a time scale and $\psi_0(t)$ is some function decaying sufficiently fast on both sides of its single maximum; further on, $\psi_0(t) = \exp(-t^2)$ shall be used. The wavelet function is defined as: $\psi_n(t) = (-1)^n \cdot \frac{d^n\psi_0(t)}{dt^n} \equiv (-1)^n \cdot \psi^{(n)}{}_0(t)$. Using integration by part and exploiting the fast decay of the function $\psi_0(t)$ at $t \to \pm\infty$, the following formula for the Taylor's coefficients (the n-th derivative of the smoothed signal, divided by $n!$) of the smoothed signal is obtained:

$$c_n(t,a) \equiv \frac{1}{n!}\frac{d^n c_0(t,a)}{dt^n} = \int_{-\infty}^{+\infty} Y(t+av)\psi_n(v)\,dv / a^n \int_{-\infty}^{+\infty} v^n\psi_n(v)\,dv \tag{10}$$

The WTMM-point (t,a) for $n \geq 1$ is defined as the point for which $|c_n(t,a)|$ exhibits a local maximum with respect to time t for a given time scale a. For $n = 0$, the WTMM-points are defined as points of local extremes (maxima or minima) of the smoothed signal $c_0(t,a)$, that may be joined to form chains. The set of all chains forms a WTMM-skeleton of the signal. If $\psi_0(t)$ is a Gaussian, then a given WTMM-skeleton chain does not end when the scale a is decreased [52]. The WTMM-points for the 1st-order derivative $c_1(t,a)$ indicate time points of the maximum trend (positive or negative) of the smoothed signal $c_0(t,a)$ for the given scale. This can be used to extract points of large and abrupt changes of the mean value. These times correspond to the occurrence of rather long WTMM-chains that begin to grow. Let us define a WTMM stepwise approximation $S_Y(t|a_*)$ for the signal $Y(t)$, as a function which is equal to the sequence of constant values g_k in the successive intervals $t \in [\tau_k(a_*), \tau_{k+1}(a_*)]$. Here $\tau_k(a_*)$ are the beginnings of the WTMM-chains for $c_1(t,a)$ which exceed the threshold time scale a_* and $g_k = \sum_{t=\tau_k(a_*)}^{\tau_{k+1}(a_*)} Y(t)/(\tau_{k+1}(a_*) - \tau_k(a_*) + 1)$ is equal to the mean value of $Y(t)$ within the time interval $[\tau_k(a_*), \tau_{k+1}(a_*)]$.

Let us calculate the weighted average entropy of all clusters. As weights we use the squares of the eigenvector components of the correlation matrix of the entropy values of the first principal components of all clusters. The calculation of such an average is the following application of the principal component method. In **Figure 4b** the black line shows the graph of the weighted average entropy, and in **Figure 4c** the values of the weights. From graph 4(c), it can be seen that the weight of cluster #7 stands out significantly compared to the weights of other clusters. Thus, cluster #7 could be regarded as the most informative. The WTMM piecewise stepwise approximation for time variations of the weighted average entropy of the principal components from all clusters is represented in **Figure 4b** by the red line. To construct this approximation, a sampling scale $a_* = 10$ was used. Since the entropy values were calculated in a sliding time window of length 365 days with an offset of 3 days, the dimensional value of this scale is equal to 30 days. The absolute values of the increments of the average entropy for the piecewise-step approximation in **Figure 4a** form a sequence of 56 "events" of sufficiently large changes in the average entropy.

The graphs in **Figure 4** provide an integral assessment of the change in entropy of the entire region and allow us to identify a close to periodic regime (with a period of about 3 years) of changes in the statistical structure of the earth's surface tremor. In addition, a general trend of decreasing entropy over the entire observation period is visible. A decrease in average entropy means that the composition of GPS time series is enriched with non-stationary components, which reflect an increase in the intensity of mutual movements of small blocks of the earth's crust.

Let us try to find out whether there is a statistically significant periodicity in the occurrence of jumps in the average entropy value, presented in **Figure 4a**. To test this assumption, let us estimate the periodic components of the shock intensity using the parametric model proposed in [53].

Let $t_i,\ i = 1, \dots, N$ be the times of the sequence of events observed on the interval $(0, T]$. Consider the following intensity model containing a periodic component:

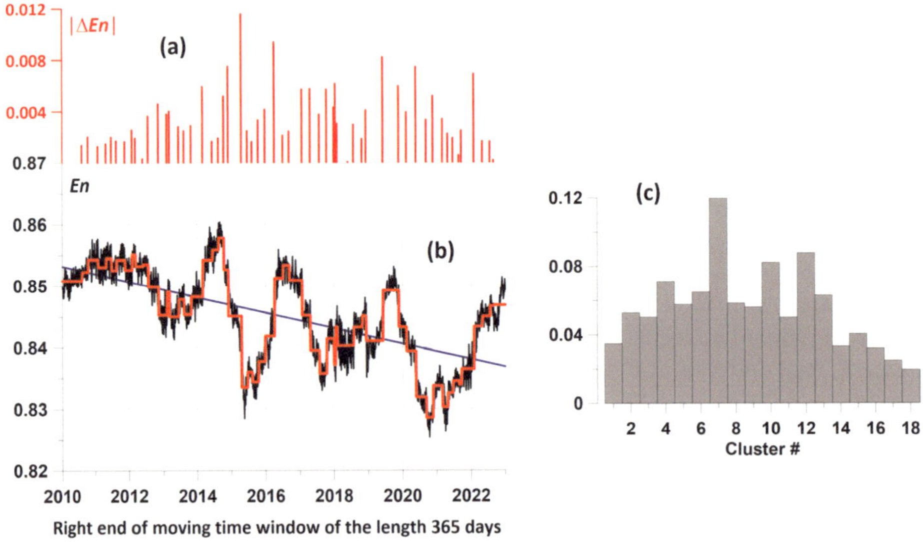

Figure 4.
(a) – absolute increments of average weighted entropy (b, black line) for piecewise stepwise approximation (b, red line). The blue line in plot (b) represents the linear trend of the average entropy. Figure (c) represents the weight values of each cluster as the squares of the eigenvector components corresponding to the maximum eigenvalue of the correlation matrix.

$$\lambda(t) = \mu \cdot (1 + a\cos(\omega t + \varphi)) \tag{11}$$

where frequency ω, amplitude a, $0 \le a \le 1$, phase angle φ, $\varphi \in [0, 2\pi]$ and multiplier $\mu > 0$ (describing the Poisson part of the intensity) are the model parameters. Thus, the Poisson part of the intensity is modulated by a harmonic oscillation. Let us fix some frequency value ω. The log likelihood function [54] in this case for a series of observed events is equal to:

$$\begin{aligned}\ln L(\mu, a, \varphi|\omega) &= \sum_{t_i} \ln(\lambda(t_i)) - \int_0^T \lambda(u)du = \\ &= N\ln(\mu) + \sum_{t_i} \ln(1 + a\cos(\omega t_i + \varphi)) - \mu T - \frac{\mu a}{\omega}[\sin(\omega T + \varphi) - \sin(\varphi)]\end{aligned} \tag{12}$$

Taking the maximum of expression Eq. (12) with respect to the parameter μ, it is easy to find that:

$$\mu = \hat{\mu}(a, \varphi|\omega) = \frac{N}{T + a(\sin(\omega T + \varphi) - \sin(\varphi))/\omega} \tag{13}$$

Substituting Eq. (13) into formula Eq. (12) we get

$$\ln(L(\hat{\mu}, a, \varphi|\omega)) = \sum_{t_i} \ln(1 + a\cos(\omega t_i + \varphi)) + N \cdot \ln(\hat{\mu}(a, \varphi|\omega)) - N \tag{14}$$

It should be noted that the expression $\hat{\mu}(a = 0, \varphi|\omega) \equiv \hat{\mu}_0 = N/T$ is an estimate of the intensity of the process, provided that it is homogeneous Poisson (purely random). Thus, the increment of the log-likelihood function due to consideration of a richer intensity model with a harmonic component with a given frequency ω than for a purely random flow of events is equal to:

$$\Delta \ln L(a, \varphi|\omega) = \sum_{t_i} \ln(1 + a\cos(\omega t_i + \varphi)) + N \cdot \ln(\hat{\mu}(a, \varphi|\omega)/\hat{\mu}_0) \tag{15}$$

Let

$$R(\omega) = \max_{a,\varphi} \Delta \ln L(a, \varphi|\omega), \quad 0 \le a \le 1, \ \varphi \in [0, 2\pi] \tag{16}$$

Function Eq. (16) can be considered as a generalization of the spectrum for a sequence of events. The graph of this function shows how "more profitable" the periodic intensity model is compared to a purely random model. The maximum values of function Eq. (16) highlight the frequencies present in the stream of events.

An important issue in applying this method to real data is determining the statistical significance of the obtained peak values of statistics $R(\omega)$. To solve it, one can apply Wilks' asymptotic theory [55]. Let us consider 2 hypotheses for the same data set consisting of independent observations:

1. hypothesis H_0: $X^{(N)}$ distributed according to density $p_0(X^{(N)}|\theta_0)$;

2. hypothesis H_1: $X^{(N)}$ distributed according to density $p_1(X^{(N)}|\theta_1)$.

where θ_0 and θ_1 are vectors of unknown parameters having dimensions m_0 and m_1, and the hypothesis H_1 is more "rich": $m_1 > m_0$, and the vector of parameters θ_1 completely includes the components of the vector θ_0. Let us consider the difference between the log likelihoods for these two hypotheses, provided that the parameter vectors are taken from their maximum likelihood estimates:

$$\Delta \ln L\left(X^{(N)}\right) = \ln\left(\max_{\theta_1} p_1\left(X^{(N)}|\theta_1\right)\right) - \ln\left(\max_{\theta_0} p_0\left(X^{(N)}|\theta_0\right)\right) \quad (17)$$

It is obvious that $\Delta \ln L(X^{(N)}) \geq 0$. According to Wilks' theorem, if the hypothesis H_0 is valid, quantity Eq. (17) has an asymptotic distribution:

$$\Delta \ln L\left(X^{(N)}\right) \sim \frac{\chi_m^2}{2}, \quad m = m_1 - m_0, \quad N \to \infty \quad (18)$$

In our case $m = 2$, and therefore, the doubled value Eq. (18) has an asymptotic distribution χ_2^2 density equal to $e^{-x/2}/2$, and the value Eq. (18) itself is distributed asymptotically with an exponential density e^{-x} or

$$\Pr\{R(\omega) < x\} = 1 - e^{-x}, \quad N \to \infty \quad (19)$$

provided that the analyzed sequence of time points is distributed according to the Poisson law with constant intensity. Expression Eq. (19) allows us to give thresholds for statistics $R(\omega)$, which allow us to state that only when exceeded, the sequence of time instants differs from the Poisson sequence with a given probability. In particular, setting the threshold probability equal to 0.99, we obtain that the threshold is equal to $\ln(100) = 4.6$.

Figure 5 shows graphs of function estimates $R(\omega)$ for 1000 frequency values with periods from 10 to 1000 days, the values of which are taken uniformly on a logarithmic scale.

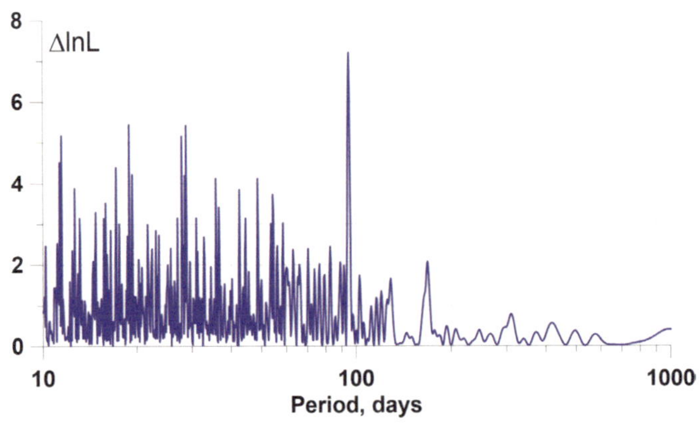

Figure 5.
Graph of increments $R(\omega)$ of the maximum value of the log-likelihood function of a periodic model of the intensity of mean entropy jumps compared to a purely random Poisson process model.

If we choose the value 4.6 as the statistical $R(\omega)$ threshold, above which model (11) has an advantage over the Poisson model with a probability exceeding 0.99, then 5 selected period values arise: 11.4, 18.9, 27.8, 28.7 and 95 days, of which the last period is clearly dominant.

6. Dynamics of extreme entropy values

Figure 6(a1,b1) shows graphs of changes in cluster numbers in which the minimum (a1) and maximum (b1) entropy values were realized for each position of a time window of 365 days in length. These graphs show that there are quite long periods of time when extreme entropy values are delayed in close clusters of stations. Having constructed histograms of the distribution of cluster numbers in which extreme entropy values are realized (**Figure 6**(a2,b2)), one can notice that there are clusters that "attract" these extreme values to themselves. In particular, entropy minima are most often realized in clusters #4 and #7, and maximum values are in clusters #15 and #13. Such differences highlight "characteristic areas" of earth tremor [39, 40], which reflect differences in the dynamics of the behavior of crustal blocks.

To represent the spatiotemporal dynamics of the behavior of extreme entropy values, we calculate a histogram of the distribution of cluster numbers in which minima and maxima are realized in a sliding time window, which includes a certain number of neighboring extreme entropy values, presented in the graphs of **Figure 6**(a1,b1).

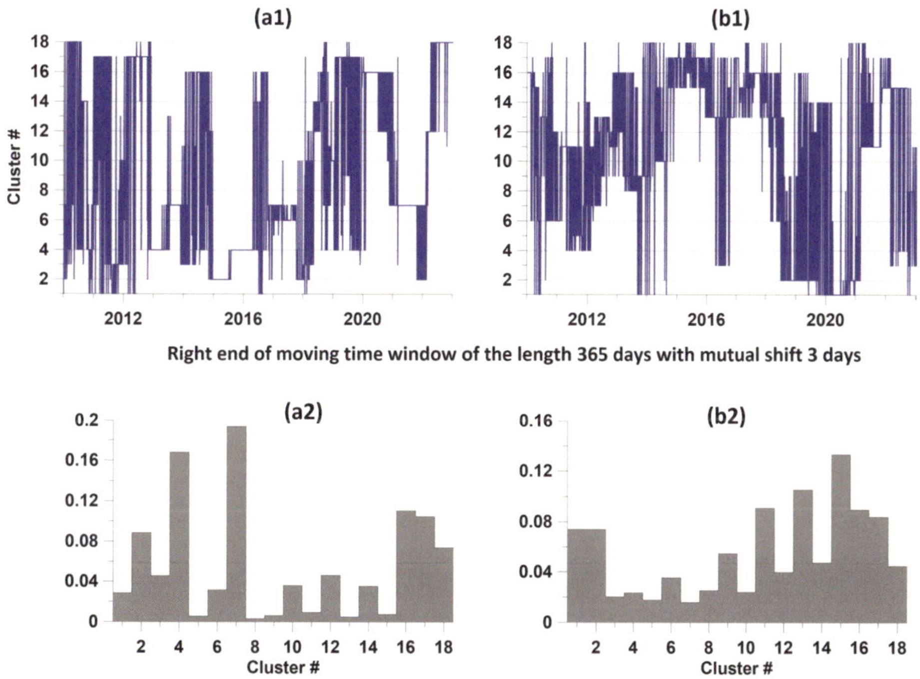

Figure 6.
Graphs (a1) and (a2) represent sequences of cluster numbers in which minima (graph (a1)) and maxima (graph (a2)) of entropy values were realized in each time window of 365 days with an offset of 3 days. Graphs (b1) and (b2) represent histograms of the distribution of cluster numbers, and which had minima (b1) and maxima (b2) of entropy.

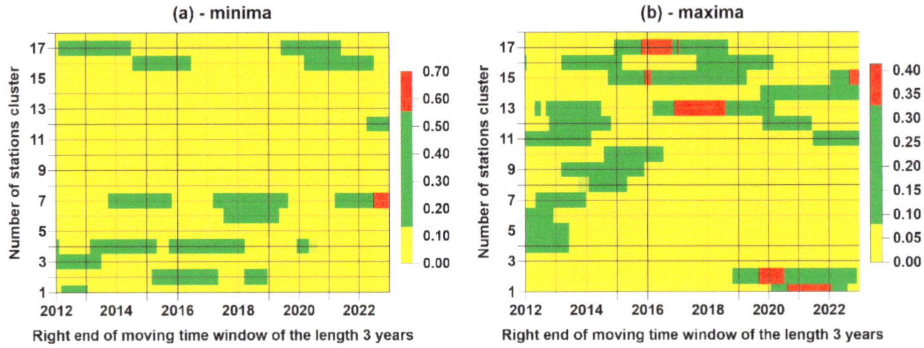

Figure 7.
Histograms of clusters numbers where minima (a) and maxima (b) of first principal component entropy are realized within moving time window of the length 3 years.

Since each entropy value is calculated in a "small" sliding time window of length 365 days with an offset of 3 days, then if we take M neighboring entropy values, then the dimensional length of the "large" time window in which the histograms are calculated will be equal to $365 + 3 \cdot (M - 1)$ days. With a value of $M = 244$, the length of the "large" time window of the histogram will be equal to 1094 days, which is approximately equal to 3 years. Histograms of extreme entropy values in a 3-year time window are presented in **Figure 7** in the form of two-dimensional diagrams. A similar method for presenting changes in the spatiotemporal properties of extreme values of global seismic noise statistics was previously used in [56].

Note that in **Figure 7b**, there is a noticeable trend in the space of cluster numbers for the evolution of the maximum values of the probability of the distribution of entropy maximums for the timestamps of the right end of the "large" 3-year time window from 2012 to 2019 from cluster #4 to cluster #17. Considering that the clusters are numbered in ascending latitude, physically this change corresponds to the migration of entropy maxima from south to north for the time interval 2009–2019. (taking into account that the time stamps in **Figure 7** correspond to the right end of the 3-year time window).

Let $p_t^{(j)}$ be the probabilities of histograms of the distribution of extreme entropy values (minimums or maximums) for the moment in time of the right end t of a "large" time window of 3 years in length, which are presented in **Figure 7**, $\sum_j p_t^{(j)} = 1$. Let us denote

$$j^*(t) = \arg \max_{1 \le j \le 18} p_t^{(j)} \tag{20}$$

the number of the cluster in which the extreme entropy value was realized in the 3-year time window with the time stamp t of the right end of the window. Let (X_j, Y_j) be the longitude and latitude of the centers of gravity of station clusters, $j = 1, \dots, 18$. Then the jump-like trajectory of changes in the centers of gravity of those cluster numbers for which the maximum probability of extreme entropy values is realized will be a sequence of points $(X_{j^*(t)}, Y_{j^*(t)})$. In addition, you can calculate the continuous weighted average trajectory of the distribution of extreme entropy values using the formulas:

$$X_t = \sum_j p^{(j)}_t \cdot X_j, \quad Y_t = \sum_j p^{(j)}_t \cdot Y_j \tag{21}$$

In **Figure 8**, the dynamics of changes in the probability distribution of extreme entropy values is transferred from the space of cluster numbers, as in **Figure 7**, to the plane of geographic coordinates. In **Figure 8a**, the blue line shows the jump trajectory $(X_{j^*(t)}, Y_{j^*(t)})$ for the distribution of minimum entropy values. In addition, in **Figure 8a**, the continuous blue line represents the weighted average trajectory calculated according to formula Eq. (21). **Figure 8b**, similar to **Figure 8a**, shows the trajectories of changes in the probabilities of the distribution of maximum entropy values. The jump-like red line corresponds to a change in time in the sequence of cluster centers $(X_{j^*(t)}, Y_{j^*(t)})$, in which for each position of the 3-year time window the maximum probability value $p^{(j)}_t$ of the histogram of the distribution of entropy maxima among all clusters of stations is realized. The continuous pink line in **Figure 8b** depicts the weighted average trajectory Eq. (21) of the maximum entropy values, which is similar to the blue line in **Figure 8a** for the minimum entropy values. Note that those clusters for which the probabilities in the histograms in **Figure 6**(a2) and **Figure 6**(b2) are small are not visited by the jump-like trajectories in **Figure 8**.

From a comparison of **Figure 8a** and **b**, it is noticeable that the region of minimum entropy values gravitates toward the Pacific coast, in particular, toward the San Andreas fault, which is apparently due to more intense movements of crustal blocks in this region.

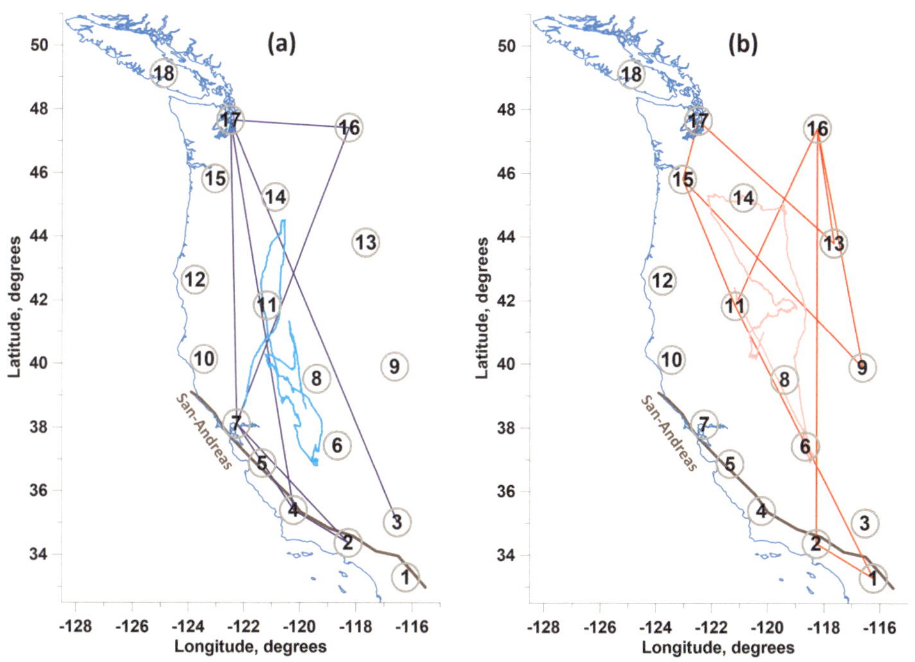

Figure 8.
*In plot (a), the blue lines show the trajectory of changes in the numbers of clusters in which the maximum probability was realized, and the cyan line shows the corresponding weighted average trajectory of the distribution of the minimum entropy values over all clusters (**Figure 5a**). In plot (b), the red lines show the trajectory of changes in the numbers of clusters in which the maximum probability was realized, and the pink line shows the corresponding weighted average trajectory of the distribution of maximum entropy values over all clusters (**Figure 5b**). The thick brown line in (a) and (b) shows the san-Andreas fault. The largest cities in California, Los Angeles and San Francisco, are located in clusters #2 and #7, respectively.*

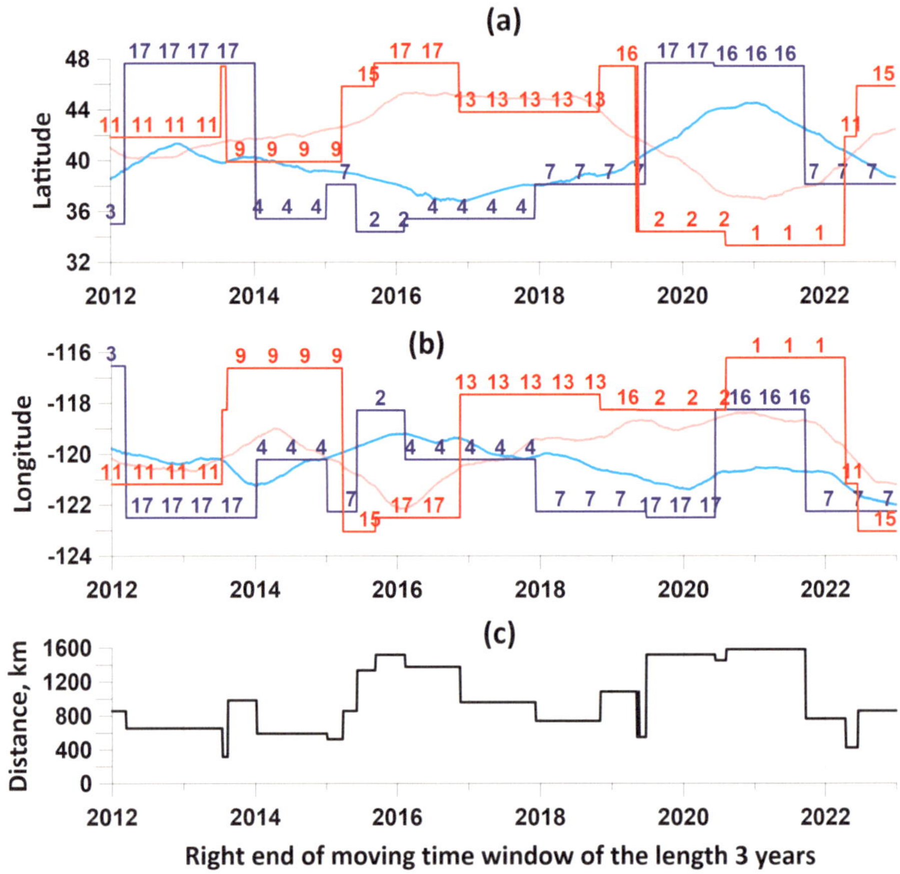

Figure 9.
In plots (a) and (b) by blue and cyan colors for minimum entropy values and by red and pink colors for maximum entropy values the changes in latitude (a) and longitude (b) are presented for maximum and weighted mean probabilities. Along the lines of changes in the coordinates of the maximum probabilities of extreme entropy values (blue for the minimum and red for the maximum), the corresponding colors mark the belonging of the trajectory sections to one or another cluster. Plot (c) presents distance between trajectories of maximum probabilities of extremum entropy values.

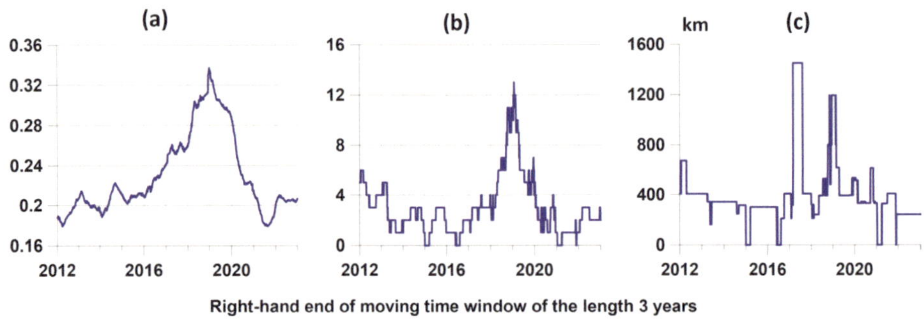

Figure 10.
(a) – mean value of absolute pair correlations between values of entropy of clusters within moving time window of the length 3 years; (b) – number of pairs of clusters for which absolute correlations exceed the threshold 0.6; (c) – maximum distance between clusters centers for which absolute correlations exceed 0.6.

Another way to present trajectories of changes in the probabilities of extreme entropy values is presented in **Figure 9**, where the trajectories are plotted separately for latitude and longitude. For jump-like trajectories of changes in maximum probabilities along piecewise-constant graphs, numerical labels are placed in the appropriate color to indicate membership in one or another cluster of stations. In **Figure 9a**, on a continuous graph of changes in the weighted average probability (pink line), one can see the trend of migration of the probability of maximum entropy values to the north (in the direction of increasing latitude) for time stamps of the right end of the 3-year window until 2019, which is noticeable in **Figure 7b**. Some clusters of stations are visited by both the trajectory of maximum probabilities of minimum entropy values (blue line in **Figure 8a**) and the trajectory of maximum probabilities of maximum entropy values (red line in **Figure 8b**). However, these trajectories never coincide in time, as follows from the graph in **Figure 9c**, which shows the distance between these trajectories.

Based on the interpretation of the entropy of GPS time series as a measure of non-stationarity, from the graphs in **Figure 9** we can conclude that for the last time fragment of 3 years in length, 2018–2023, the most “restless”, with the lowest entropy, is cluster #7, in which San-Francisco is located.

7. Spatiotemporal correlations of entropy values in station clusters

In a 3-year sliding time window, we calculate absolute values of all pairwise correlation coefficients between entropy values in each cluster. For 18 clusters, the number of such pairwise correlations is 153. **Figure 10a** shows a graph of the change in the average value of absolute pairwise correlations depending on the position of the right end of the sliding 3-year window. This graph highlights the 2017–2019 time interval as the time slice of maximum average correlations. Let us calculate the number of pairs of clusters between which an absolute correlation has arisen that exceeds the threshold of 0.6. The change in this number of pairs of clusters with the strongest correlations of entropy values is presented in **Figure 10b**, which also

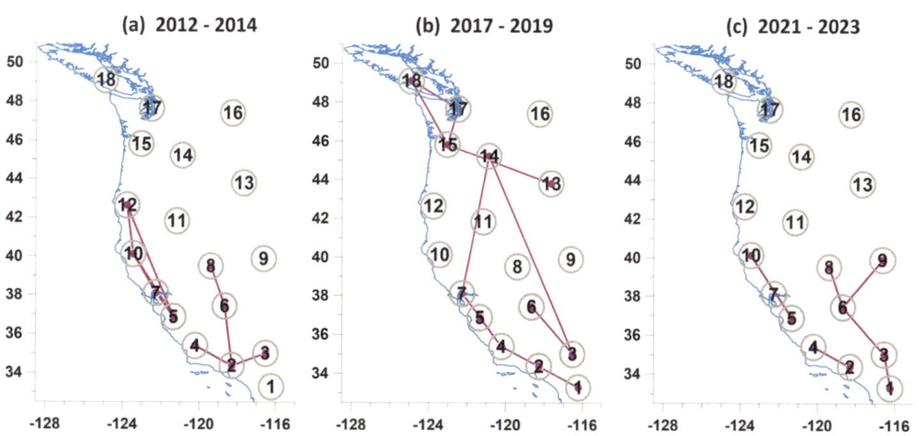

Figure 11.
Purple straight lines connect the centers of those clusters for which pairwise absolute correlations between entropy values in sliding time windows of 3 years in length exceed 0.6 for three intervals of timestamp values at the right end of the time windows.

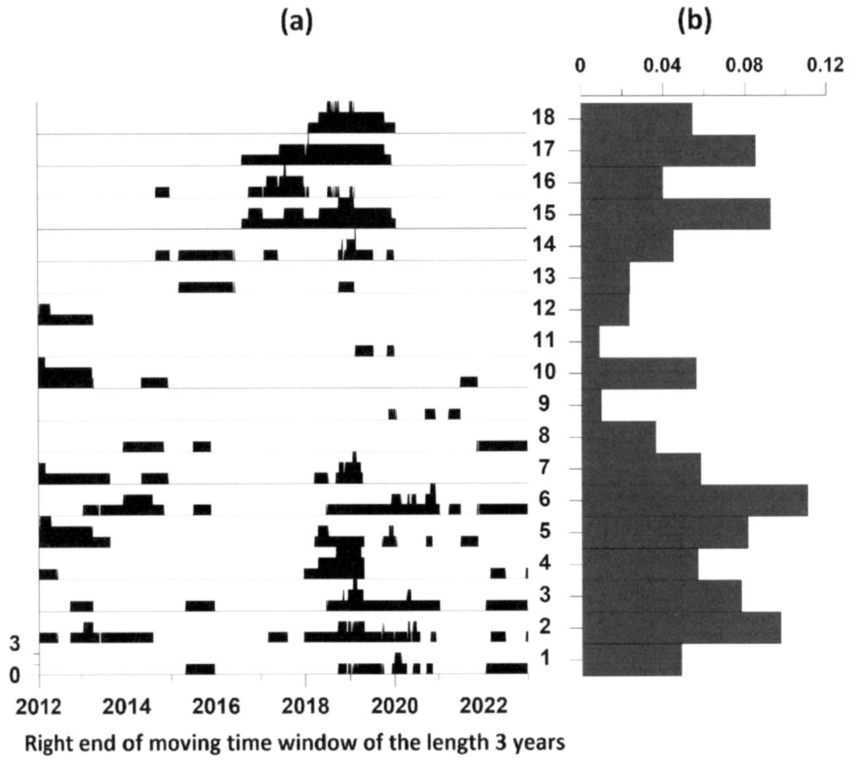

Figure 12.
Figure (a) for each cluster of stations shows graphs of the numbers of pairwise correlations of the principal components of this cluster with the principal components of other clusters exceeding the threshold of 0.6 in a sliding time window of 3 years. Figure (b) plots sample estimates of the probabilities that the correlations of the principal component of a given cluster with other clusters exceed the 0.6 threshold.

highlights the time interval 2017–2019 with the largest number of pairwise strong correlations. Finally, **Figure 10c** shows a plot of changes in the maximum distance between cluster centers for which the absolute entropy correlation exceeded 0.6. This graph contains two bursts of maximum distance, one of which (the second in a row) again falls on the time interval 2017–2019.

In **Figure 11**, to visualize strong spatial correlations, straight lines connect the centers of those clusters for which the absolute values of correlations between entropy values in a sliding time window of 3 years length exceed the threshold of 0.6. The connection of cluster centers is carried out for three intervals of timestamp values at the right end of the 3-year time window. The time stamp boundaries are chosen so that each graph in **Figure 11** corresponds to a time interval of the same length. **Figure 11a** and **c** correspond to the beginning and end of the analyzed time period. The central **Figure 11b** presents the strong correlations for the time interval corresponding to the maximum point of the average correlations and the numbers of pairs of clusters with strong correlations in **Figure 10a** and **b**. It can be seen from **Figure 11b** that strong correlations form a network that covers the entire region, while for weak average correlations in **Figure 11a** and **c** strong pairwise correlations are observed only between the southern clusters of stations and those located along the San-Andreas fault.

Figure 12 provides a detailed view of the time evolution of correlations between clusters. For each cluster, **Figure 12a** shows graphs of changes in the numbers of pairs of other clusters for which the correlation between entropy values in a sliding time window of 3 years exceeded 0.6. The maximum number of such pairwise correlations is 3 - therefore, all Y-axes in these graphs have the same scale of 3. For each cluster, **Figure 12b** shows the probability that the pairwise correlation coefficient for entropy values with other clusters will exceed 0.6. From this graph, it is clear that clusters #9 and #11 are the least correlated with other clusters and act as a kind of border dividing the entire region into northern and southern parts. **Figures 11** and **12** show that strong correlations are most often observed for cluster pairs located along the San Andreas Fault and on the Pacific Coast.

8. Conclusion

A method for joint analysis of a large number of synchronous monitoring time series is proposed using the example of data on daily vertical displacements of the earth's surface in the Western USA (California) for the observation interval 2009–2023. The analysis is based on dividing the initial data (699 GPS time series) into blocks (clusters of stations) and sequentially applying the principal component method. The minimum normalized entropy of the distribution of squares of orthogonal wavelet coefficients, which is a measure of the information content of the distribution of oscillation energy over a system of independent time-frequency intervals, was chosen as the main research tool. Particular attention is paid to time intervals and areas (clusters) with extreme entropy values. Periodicities in the occurrence of strong jumps in the average entropy of the entire region have been discovered, of which the period of 95 days is dominant. The trend of migration of areas of maximum entropy from the South to the North has been identified. As a result of the analysis, it was found that the area of minimum entropy values gravitates toward the San Andreas fault, and the vicinity of San Francisco (cluster #7) has the selected properties of maximum information content and the last 3-year time fragment (2021–2023) of low entropy trajectories belongs to this cluster.

Further development of methods for analyzing the fine structure of GPS time series, going beyond the traditional analysis of the field of displacements of the earth's surface, involves, in particular, a more detailed analysis of spectral properties. In particular, the transition from studying the field of maximum traditional Pearson correlations to the field of frequency-dependent correlations, that is, to the analysis of maximum coherences, seems promising. In this direction, it is also possible to use Hilbert-Huang expansions [57] in terms of a system of independent intrinsic vibration modes, which provides additional opportunities for studying nonlinear nonstationary effects.

Acknowledgements

The work was carried out within the framework of the state assignment of the Institute of Physics of the Earth of the Russian Academy of Sciences (topic FMWU-2022-0018).

Funding

This research received no external funding.

Conflict of interest

The authors declare no conflict of interest.

Author details

Alexey Lyubushin
Institute of Physics of the Earth RAS, Moscow, Russia

*Address all correspondence to: lyubushin@yandex.ru

References

[1] Roncagliolo PA, Garcнa JG, Mercader PI, Fuhrmann DR, Muravchik CH. Maximum-likelihood attitude estimation using GPS signals. Digital Signal Processing. 2007;**17**: 1089-1100. DOI: 10.1016/j.dsp.2006.09.001

[2] Wang F, Li H, Lu M. GNSS spoofing detection and mitigation based on maximum likelihood estimation. Sensors. 2017;**17**:1532. DOI: 10.3390/s17071532

[3] Parkinson BW. Global Positioning System: Theory and Applications. Reston, VA, USA: AIAA; 1996 781 p

[4] Jing X, He J, Zhang Y, Fei X, Cai F. A distance-based maximum likelihood estimation method for sensor localization in wireless sensor networks. International Journal of Distributed Sensor Networks. 2016;**12**(4). Article ID 2080536. DOI: 10.1155/2016/2080536

[5] Langbein J, Johnson H. Correlated errors in geodetic time series, implications for time-dependent deformation. Journal of Geophysical Research. 1997;**102**(B1):591-603. DOI: 10.1029/96JB02945

[6] Williams SDP, Bock Y, Fang P, Jamason P, Nikolaidis RM, Prawirodirdjo L, et al. Error analysis of continuous GPS time series. Journal of Geophysical Research. 2004;**109**(B3): B03412

[7] Bos MS, Fernandes RMS, Williams SDP, Bastos L. Fast error analysis of continuous GPS observations. Journal of Geodesy. 2008;**82**(3):157-166. DOI: 10.1007/s00190-007-0165-x

[8] Wang W, Zhao B, Wang Q, Yang S. Noise analysis of continuous GPS coordinate time series for CMONOC. Advances in Space Research. 2012;**49**(5): 943-956. DOI: 10.1016/j.asr.2011.11.032

[9] Agnew D. The time domain behavior of power law noises. Geophysical Research Letters. 1992;**19**:333-336. DOI: 10.1029/91GL02832

[10] Amiri-Simkooei AR, Tiberius CCJM, Teunissen PJG. Assessment of noise in GPS coordinate time series: Methodology and results. Journal of Geophysical Research. 2007;**112**:B07413. DOI: 10.1029/2006JB004913

[11] Caporali A. Average strain rate in the Italian crust inferred from a permanent GPS network – I. Statistical analysis of the time-series of permanent GPS stations. Geophysical Journal International. 2003;**155**:241-253. DOI: 10.1046/j.1365-246X.2003.02034.x

[12] Zhang J, Bock Y, Johnson H, Fang P, Williams S, Genrich J, et al. Southern California permanent GPS geodetic array: Error analysis of daily position estimates and site velocities. Journal of Geophysical Research. 1997;**102**(B8): 18,035-18,055. DOI: 10.1029/97JB01380

[13] Li J, Miyashita K, Kato T, Miyazaki S. GPS time series modeling by autoregressive moving average method, application to the crustal deformation in Central Japan. Earth, Planets and Space. 2000;**52**:155-162. DOI: 10.1186/BF03351624

[14] Kermarrec G, Schon S. On modelling GPS phase correlations: A parametric model. Acta Geodaetica et Geophysica. 2018;**53**:139-156. DOI: 10.1007/s40328-017-0209-5

[15] Beavan J. Noise properties of continuous GPS data from concrete

pillar geodetic monuments in New Zealand and comparison with data from U.S. deep drilled braced monuments. Journal of Geophysical Research. 2005; **110**:B08. DOI: 10.1029/2005JB003642

[16] Langbein J. Noise in GPS displacement measurements from Southern California and southern Nevada. Journal of Geophysical Research. 2008;**113**:B05405. DOI: 10.1029/2007JB005247

[17] Blewitt G, Lavallee D. Effects of annual signal on geodetic velocity. Journal of Geophysical Research. 2002; **107**(B7):2145. DOI: 10.1029/2001JB000570

[18] Bos MS, Bastos L, Fernandes RMS. The influence of seasonal signals on the estimation of the tectonic motion in short continuous GPS time-series. Journal of Geodynamics. 2010;**49**(3–4): 205-209. DOI: 10.1016/j.jog.2009.10.005

[19] Liu B, Xing X, Tan J, Xia Q. Modeling seasonal variations in vertical GPS coordinate time series using independent component analysis and varying coefficient regression. Sensors. 2020;**20**:5627. DOI: 10.3390/s20195627

[20] Tesmer V, Steigenberger P, van Dam T, Mayer-Gurr T. Vertical deformations from homogeneously processed GRACE and global GPS long-term series. Journal of Geodesy. 2011;**85**: 291-310. DOI: 10.1007/s00190-010-0437-8

[21] Yan J, Dong D, Burgmann R, Materna K, Tan W, Peng Y, et al. Separation of sources of seasonal uplift in China using independent component analysis of GNSS time series. Journal of Geophysical Research - Solid Earth. 2019;**124**:11951-11971. DOI: 10.1029/2019JB018139

[22] Pan Y, Shen W-B, Ding H, Hwang C, Li J, Zhang T. The quasi-biennial vertical oscillations at global GPS stations: Identification by ensemble empirical mode decomposition. Sensors. 2015;**15**: 26096-26114. DOI: 10.3390/s151026096

[23] Liu B, Dai W, Liu N. Extracting seasonal deformations of the Nepal Himalaya region from vertical GPS position time series using independent component analysis. Advances in Space Research. 2017;**60**:2910-2917. DOI: 10.1016/j.asr.2017.02.028

[24] Santamarнa-Gymez A, Memin A. Geodetic secular velocity errors due to interannual surface loading deformation. Geophysical Journal International. 2015; **202**:763-767. DOI: 10.1093/gji/ggv190

[25] Fu Y, Argus DF, Freymueller JT, Heflin MB. Horizontal motion in elastic response to seasonal loading of rain water in the Amazon Basin and monsoon water in Southeast Asia observed by GPS and inferred from GRACE. Geophysical Research Letters. 2013;**40**:6048-6053. DOI: 10.1002/2013GL058093

[26] Chanard K, Avouac JP, Ramillien G, Genrich J. Modeling deformation induced by seasonal variations of continental water in the Himalaya region: Sensitivity to earth elastic structure. Journal of Geophysical Research - Solid Earth. 2014;**119**: 5097-5113. DOI: 10.1002/2013JB010451

[27] He X, Montillet JP, Fernandes R, Bos M, Yu K, Hua X, et al. Review of current GPS methodologies for producing accurate time series and their error sources. Journal of Geodynamics. 2017;**106**:12-29. DOI: 10.1016/j.jog.2017.01.004

[28] Ray J, Altamimi Z, Collilieux X, van Dam T. Anomalous harmonics in the spectra of GPS position estimates. GPS

Solutions. 2008;**12**:55-64. DOI: 10.1007/s10291-007-0067-7

[29] Teferle FN, Williams SDP, Kierulf HP, Bingley RM, Plag HP. A continuous GPS coordinate time series analysis strategy for high-accuracy vertical land movements. Physics Chemistry Earth, Parts A/B/C. 2008;**33** (3–4):205-216. DOI: 10.1016/j.pce.2006.11.002

[30] Chen Q, van Dam T, Sneeuw N, Collilieux X, Weigelt M, Rebischung P. Singular spectrum analysis for modeling seasonal signals from GPS time series. Journal of Geodynamics. 2013;**72**:25-35. DOI: 10.1016/j.jog.2013.05.005

[31] Bock Y, Melgar D, Crowell B.W. Real-time strong-motion broadband displacements from collocated GPS and accelerometers, Bulletin of the Seismological Society of America. 2011;**101**(6):2904-2925. 10.1785/0120110007

[32] Hackl M, Malservisi R, Hugentobler U, Jiang Y. Velocity covariance in the presence of anisotropic time correlated noise and transient events in GPS time series. Journal of Geodynamics. 2013;**72**:36-45. DOI: 10.1016/j.jog.2013.08.007

[33] Goudarzi MA, Cocard M, Santerre R, Woldai T. GPS interactive time series analysis software. GPS Solutions. 2013; **17**(4):595-603. DOI: 10.1007/s10291-012-0296-2

[34] Khelif S, Kahlouche S, Belbachir MF. Analysis of position time series of GPS-DORIS co-located stations. International Journal of Applied Earth and Observational Geointerface. 2013;**20**: 67-76. DOI: 10.1016/j.jag.2011.12.011

[35] Lyubushin A. Global coherence of GPS-measured high-frequency surface tremor motions. GPS Solutions. 2018;**22**: 116. DOI: 10.1007/s10291-018-0781-3

[36] Lyubushin A. Field of coherence of GPS-measured earth tremors. GPS Solutions. 2019;**23**:120. DOI: 10.1007/s10291-019-0909-0

[37] Filatov DM, Lyubushin AA. Fractal analysis of GPS time series for early detection of disastrous seismic events. Physica A. 2017;**469**(1):718-730. DOI: 10.1016/j.physa.2016.11.046

[38] Filatov DM, Lyubushin AA. Precursory analysis of GPS time series for seismic Hazard assessment. Pure and Applied Geophysics. 2019;**177**(1): 509-530. DOI: 10.1007/s00024-018-2079-3

[39] Lyubushin A. Identification of areas of anomalous tremor of the Earth's surface on the Japanese Islands according to GPS data. Applied Sciences. 2022;**12**:7297. DOI: 10.3390/app12147297

[40] Lyubushin A. Singular points of the tremor of the Earth's surface. Applied Sciences. 2023;**13**:10060. DOI: 10.3390/app131810060

[41] Lyubushin AA. Seismic noise wavelet-based entropy in Southern California. Journal of Seismology. 2021; **2021**(25):25-39. DOI: 10.1007/s10950-020-09950-3

[42] Blewitt G, Hammond WC, Kreemer C. Harnessing the GPS data explosion for interdisciplinary science. Eos. No.99. 2018. 10.1029/2018EO104623

[43] Duda RO, Hart PE, Stork DG. Pattern Classification. New York, Chichester, Brisbane, Singapore, Toronto: Wiley-Interscience Publication; 2000. p. 738

[44] Vogel MA, Wong AKC. PFS clustering method. IEEE Transactions on Pattern Analysis and Machine Intelligence. 1979;**1**(3):237-245. DOI: 10.1109/tpami.1979.4766919. Available from: https://pubmed.ncbi.nlm.nih.gov/21868854/

[45] Jolliffe IT. Principal Component Analysis. New York, NY: Springer-Verlag; 1986. 519 p. DOI: 10.1007/b98835

[46] Mallat SA. Wavelet Tour of Signal Processing. 2nd ed. San Diego, London, Boston, New York, Sydney, Tokyo, Toronto: Academic Press; 1999. 637 p

[47] Lyubushin A. Low-frequency seismic noise properties in the Japanese Islands. Entropy. 2021;**23**:474. DOI: 10.3390/e23040474

[48] Costa M, Goldberger AL, Peng C-K. Multiscale entropy analysis of biological signals. Physical Review. 2005;**71**(2 Pt 1): 021906. DOI: 10.1103/PhysRevE.71.021906

[49] Vallianatos F, Koutalonis I, Chatzopoulos G. Evidence of Tsallis entropy signature on medicane induced ambient seismic signals. Physica A: Statistical Mechanics and its Applications. 2019;**520**:35-43. DOI: 10.1016/j.physa.2018.12.045

[50] Varotsos PA, Sarlis NV, Skordas ES, Lazaridou MS. Entropy in the natural time domain. Physical Review. 2004;**70**: 011106. DOI: 10.1103/PhysRevE.70.011106

[51] Mallat S, Zhong S. Characterization of signals from multiscale edges. IEEE Transactions on Pattern Recognition and Machine Intellegience. 1992;**14**:710-732. DOI: 10.1109/34.142909

[52] Hummel B, Moniot R. Reconstruction from zero-crossings in scale-space. IEEE Transactions on Acoustics, Speech, and Signal Processing. 1989;**37**:2111-2130. DOI: 10.1109/29.45555

[53] Lyubushin AA, Pisarenko VF, Ruzich VV, Buddo VY. A new method for identifying seismicity periodicities. Volcanology and Seismology. 1998;**20**: 73-89

[54] Cox DR, Lewis PAW. The statistical Analysis of Series of Events. London: Methuen; 1966 285 p

[55] Rao CR. Linear Statistical Inference and Its Applications. N.Y., London, Sydney: John Wiley & Sons, Inc; 1965 522 p

[56] Lyubushin A. Spatial correlations of global seismic noise properties. Applied Sciences. 2023;**13**(12):6958. DOI: 10.3390/app13126958

[57] Huang NE, Shen Z, Long SR, Wu VC, Shih HH, Zheng Q, et al. The empirical mode decomposition and the Hilbert spectrum for nonlinear and non-stationary time series analysis. Proceedings of the Royal Society. 1998; **454**:903-995. DOI: 10.1098/rspa.1998.0193

Chapter 10

Space and Ground-Based Earth Observations for Hazards Valuation and Disaster Risk Management in the Era of Climate Change

Nasradeen A.H. Gadallah and Ahmed A.H. Siddig

Abstract

Quantifying hazards and assessing the risks in the era of climate change using Space and Ground-based Earth Observations (SAGEOs) is playing a key role in facilitating the implementation of frameworks and are essential for observing and assessing how risks have changed in recent years, as well as tracking the reduction in the level of exposure of communities to the hazards. SGEOs provide the context, scale, and perspective needed to understand various hazards, such as floods, seismic activities, wildfire, and coastal erosion, and are crucial for informing risk reduction and disaster management efforts. Correspondingly, SAGEOs contribute to the development of early warning systems for climate-related hazards, supporting timely and effective disaster preparedness and response. The integration of SAGEOs supports risk-informed decision-making by providing essential information for vulnerability and exposure mapping, thus informing adaptation planning and the development of climate-resilient strategies. In conclusion, quantifying hazards and assessing the risks of climate change using SAGEOs is vital for understanding, monitoring, and mitigating the impacts of climate-related hazards. These observations provide valuable data for characterizing hazards, developing early warning systems, and informing risk reduction and disaster management efforts, thus playing a critical role in building climate resilience and reducing disaster risks.

Keywords: remote sensing, geographic information systems, hazard assessment, community-based disaster risk management, vulnerability assessment

1. Introduction

Climate change in these eras has led tremendously to an increase in the frequency and severity of climate-related disasters and hazards, posing massive risks to human populations and the environment [1–3]. Climate change simply symbolizes long-term changes in the average weather patterns and it incorporates various disruptive impacts, including extreme weather events, sea level rise, droughts and water scarcity, wildfires, melting glaciers, heatwaves, changes in ecosystems, ocean acidification, vector-borne diseases, and agricultural impacts [4–6]. The terminologies associated with climate-related disasters and hazards consist of exposure, vulnerability, disaster, risk, resilience, and adaptive capacity.

Exposure refers to the presence of people, infrastructure, or assets in areas that are susceptible to the impacts of hazards, while vulnerability donates the characteristics and circumstances that make a community or system susceptible to the damaging effects of a hazard. A disaster is a sudden, catastrophic event that causes significant and widespread destruction, disruption, and distress, often resulting in a need for external assistance, which can take various forms and may be natural or human-made. Risk is the potential for a hazard to cause harmful consequences, and it is influenced by the interaction of exposure, vulnerability, and the severity of the hazard [1]. Resilience is the ability of a system or community to absorb and recover from the impacts of a hazard, while adaptive capacity is the potential for adjustment and transformation in response to changing conditions. One way to address these issues effectively is by formulating and enacting disaster risk reduction strategies through accurate geographic, weather, and climate information, vulnerability assessments, and early warning systems to mitigate the negative impact of extreme weather events and build resilience in the face of climatic hazards [7, 8]. This chapter is a comprehensive guide to leveraging observational insights for informed decision-making and strategic disaster preparedness.

2. Understanding hazards and climate change

2.1 Climate change hazards and their significant impacts

Climate change direct or indirectly driving a wide array of events that pose significant challenges to human populations and the environment [9–11]. Climate change and hazard and disaster risk cover the various natural, biological, technological, or anthropogenic phenomena that can trigger disasters and expose vulnerabilities within human populations and ecosystems [12]. Understanding the types of hazards is crucial for effective disaster risk reduction and climate change adaptation measures [7, 11]. For instance, the following illustrations showcasing these events and their significant impacts on humans and environment:

- *Extreme weather events*: These include heavy rainfall, floods, cyclones, and droughts, which have increased in frequency, intensity, and severity due to anthropogenic warming of the earth's atmosphere and oceans [11].

- *Sea level rise:* Rising sea levels cause a significant threat to coastal communities and ecosystems, increasing the risk of flooding and erosion [9].

- *Droughts and water scarcity*: Climate change has led to an increase in the frequency and severity of droughts, which can have significant impacts on agriculture, water resources, and human populations [13].

- *Wildfires:* Climate change has driven the frequency and intensity of wildfires, which had devastating effects on ecosystems and human populations [14].

- *Melting glaciers*: Melting glaciers can lead to increased flooding, erosion, and changes in water resources and ecosystems [9].

- *Heatwaves:* Climate change has contributed to the frequency and intensity of heatwaves that have significant impacts on human health and well-being [1].

- *Changes in ecosystems*: Many studies showed that climate change has led to huge deviations in ecosystems, including shifts in species distributions, changes in ecosystem services, and increased vulnerability to the hazards [1].

- *Ocean acidification*: Climate change has impacted oceans by acidification, that affected marine ecosystems and the services they provide [1].

- *Vector-borne diseases*: Plethora of literature reported that climate change led to an increase in the prevalence of vector-borne diseases, such as malaria and dengue fever, which can have significant impacts on human health [15, 16].

- *Agricultural impacts*: Climate change have had momentous impacts on crop yields, increased vulnerability to other hazards, and changes in the availability of water resources [11].

2.2 Frequency, distribution and severity of climate change disasters

The frequency, distribution and severity of climate change-related disasters is highly depend on the frequency and severity of extreme events leading to significant disruptions and harm to human populations and ecosystems [11]. Accordingly, the distribution of hazards and risks is influenced by various factors, including natural phenomena, socioeconomic parameters, and anthropogenic activities. Hazards, such as floods, fires, earthquakes, vary in occurrence spatially and temporally. Geographically, Africa floods, droughts, cyclones, heatwaves, and vector-borne, diseases are rampant in Africa and floods, fires, and earthquakes are predominant in Western Asia [10]. Coastal zones, experiencing high anthropic pressures, face increased exposure, vulnerability, and disaster risk. Temporally, the occurrences of climate-related disasters are varied from time to time (**Figure 1**). Understanding the distribution of these hazards and risks is essential for effective disaster risk management, as it helps identify areas of high research coverage and gaps [5].

2.3 The relationship between hazards and climate change

The relationship between climate change and hazard and disaster risk is complex and multi-faceted [1]. climate change can both directly and indirectly influence the frequency, intensity, and distribution of hazards such as extreme weather events and natural disasters [17]. In Addition, it increases the vulnerability of communities to these hazards and exacerbate the risks they face. Directly, climate change can lead to an increase in the frequency and intensity of hazards such as floods, storms, and droughts. Indirectly, climate change contributes to the degradation of ecosystems and natural resources, reducing their ability to buffer against hazards [18]. Moreover, climate change interrupt weather patterns and alter the timing and distribution of rainfall, leading to changes in water availability and exacerbating drought conditions [17]. These changes in hazard patterns and increased vulnerability can ultimately result in a higher likelihood of disasters occurring and greater impacts when they do happen. Simply, climate change can worsen the severity and frequency of natural hazards and disasters, making communities more vulnerable to their impacts [18].

correlation structure of the survey variable, y_i. The correlation $\rho_y(dist(s_i, s_{i'}))$ between two observations $(y_i, y_{i'})$ at the points of coordinates s_i and $s_{i'}$ is a positive and decreasing function of the distance, $d = dist(s_i, s_{i'})$: it decreases when the distance between these points increases. Details on how to use a land-use map to estimate the spatial correlation structure of crops data, using variogram and correlogram models [16], and on how to use this structure for sample design are in [17]. Here, we propose to use a CTM to estimate the parameters of the correlogram model. We use the estimated correlogram to evaluate the (anticipated) sampling variance [18] as a function of the design variables (sampling unit size and sample size) [19, 20].

We consider a simple case: to estimate the total of the survey variable in the population, $y_N = \sum_{i=1}^{N} y_i$, using a simple random sample (SRS) scheme (equal probability of inclusion, $\pi_i = n/N$) and the estimator $\hat{y}_{N_{SRS}} = N/n \sum_{i=1}^{n} y_i$. The sampling variance is $V\hat{y}_{N_{SRS}} = N^2(1 - n/N)S^2/n$, where and $\bar{y}_N = y_N/N$ are the variance and the mean of the survey variable in the population, respectively.

Pixels are arranged in rows and a columns, and we consider the sampling unit as a square block of n_0 pixels and identify each sampling unit $i = 1, 2, \cdots, N$ by the row and column of the pixel in the lower left corner of the square block. The distance $d = dist(s_i, s_{i'})$ can be expressed as $d = \sqrt{u^2 + v^2}$, where u is the number of sampling units between s_i and $s_{i'}$ in the row direction, and v is the number of sampling units between s_i and $s_{i'}$ in the column direction. The two most often used models to assess the structure of $\rho_y(d)$ are the exponential, $\rho(u, v|a, \tau) = (1 - \tau)e^{-d/a}$, and the spherical $\rho(u, v|a, \tau) = (1 - \tau)e^{-d/a}\left[1 - \frac{3}{2}\frac{d}{a} + \frac{d^3}{2a^3}\right]$ if $d \le a$ and $\rho(u, v|a, \tau) = 0$ if $d > a$.

The model parameters are the range rate a and ratio $\tau = \tau_0/(\tau_0 + \tau_d)$. Here, τ_0 is the nugget effect, that is, the variation at or near the origin (independent of distance); τ_d is the partial sill (a function of distance d between sampling points); and $(\tau_0 + \tau_d)$ is the sill, that is, the maximum variation far from the origin.

The anticipated variance is the expected value (based on the correlogram model) of the sampling variance $AV\hat{y}_{N_{SRS}} = EV\hat{y}_{N_{SRS}} = N^2(1 - n/N)ES^2/n$, where $ES^2 = \sigma^2\Psi(N, n_0|a, \tau)$, σ^2 is the variance of y_i, and $\Psi(N, n_0|a, \tau) = n_0(Nn_0 - 1)/(N - 1)[1 - \Phi(Nn_0|a, \tau)] - Nn_0(n_0 - 1)/(N - 1)[1 - \Phi(n_0|a, \tau)]$. Here, $\Phi(Nn_0|a, \tau)$ is the average correlation between pairs of pixels over the $C^2_{Nn_0}$ pairs in the population, and $\Phi(n_0|a, \tau)$ is the average correlation between pairs of pixels over the $C^2_{n_0}$ pairs in a sampling unit.

The sample design is optimized to find the sampling unit size, n_0, and the sample size, n, that minimize the sampling variance, $\min_{\{(n, n_0)\}} AV\hat{y}_{N_{SRS}} = \min_{\{(n, n_0)\}} N^2(1 - n/N)\sigma^2/n\Psi(N, n_0|a, \tau)$ subject to the cost constraint, $C_0 + C_w n n_0 + C_k\sqrt{An} \le C$. Here, C is the total budget, C_0 is a fixed cost independent of the sample size, C_w is the interviewing cost, C_k is a cost per distance unit, and A is the surveyed area. The solution to this problem is the optimum sampling unit size n_0 and optimum sample size n. In addition to the budget, C, this optimum solution is conditioned to the correlogram model parameters (a, τ).

5.1 Example

For illustrations, we use the case study in [20]. The data source in this case study was a land-use map, but we establish a parallelism between the map and the satellite

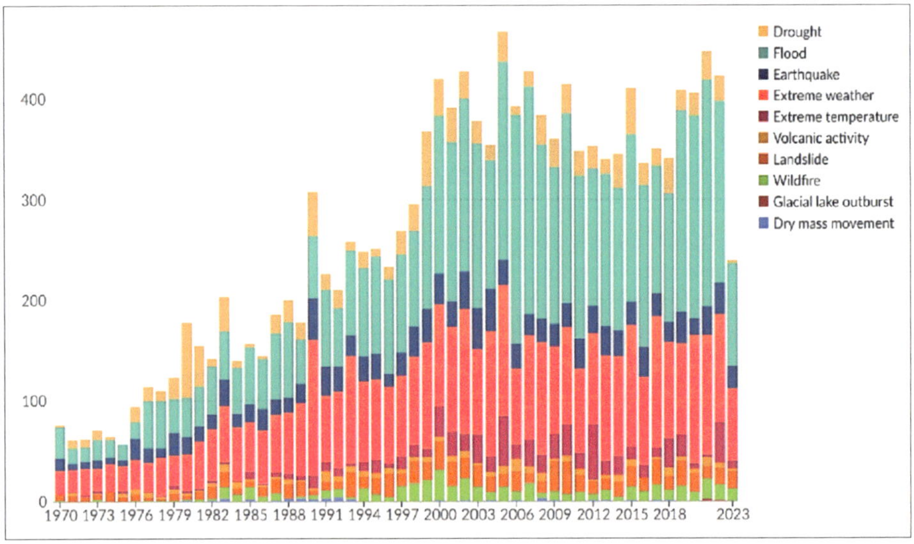

Figure 1.
Global annual reported natural disasters categorized by type, 1970 to 2023 [8]. The figure shows the frequencies and numbers of climate disaster annually, where 1975 and 2005 with least and highest occurrences respectively.

3. Space-based technologies for earth observations

3.1 Leveraging satellites to monitor hazards and climate change impacts

The 1960s has witnessed the evolution of these sources; where data from the TIROS-1 satellite delivered meteorological forecasts for the first time [19]. This was a major discovery in Earth Observation opened a new era for disaster risk management, whereby meteorological hazards can be better identified, monitored, understood and ultimately anticipated [20]. Accordingly, remarkable integration of space-based observations came true to report the coastal, hydro-meteorological and geohazards hazards such as soil and coastal erosion, groundwater and inland flooding, earthquakes and volcanic eruptions [19, 21]. The sensors are varied according to the platform and spectral characteristics (**Figure 2**).

Now, the opportunity of using satellite-based Earth observations for disaster risk assessment and management is practically more than ever [23]. These opportunities are not limited to hazard monitoring only, but it gives an opportunity for better understanding on how the vulnerability and exposure have evolved in the past, showing their current status and may evolve in the future [24]. Thus, satellites have been provide valuable information on ocean currents, atmospheric conditions, and land use patterns that are essential for managing resources and addressing climate change challenges via real-time data and imagery enabling scientists and policymakers to rake decisions to protect communities and the environment [25, 26].

3.2 Space-based satellites for climatic hazards monitoring

There are several of satellites designed for observing climatic hazards that play a crucial role in monitoring and observing various climatic hazards, helping to mitigate their impact and improve our understanding of these phenomena [13, 19, 23].

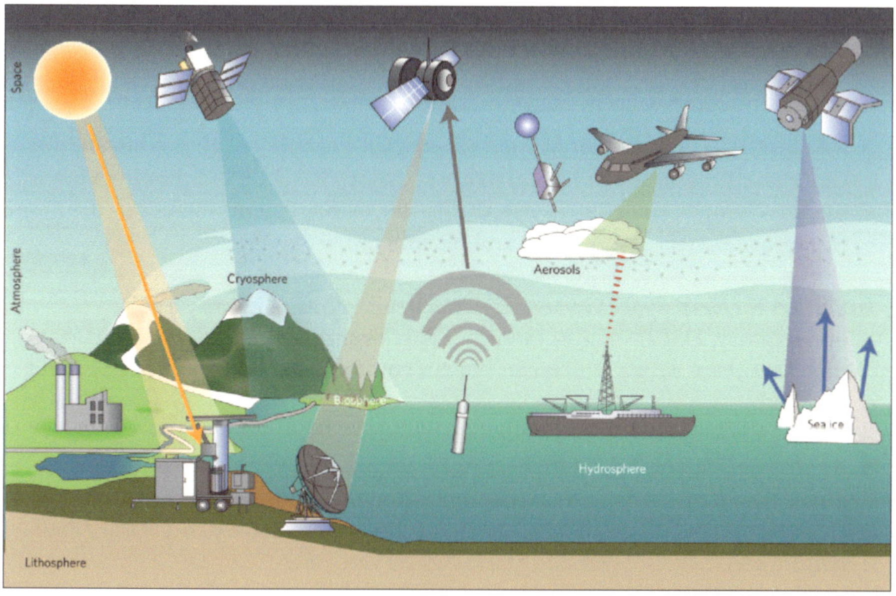

Figure 2.
Remote sensing is carried out by sensors aboard different platforms, including plane and boat. Ground-based instruments are also used, for example, sun spectral radiometers measure solar radiation [22].

These satellites differ in their sensors, resolution, coverage, and data products, making them suitable for various applications in disaster risk reduction and management [27].

The most common satellites are:

1. *Geostationary Operational Environmental Satellites (GOES) Series (NASA/NOAA):* The GOES provide continuous monitoring of weather and atmospheric conditions, including hurricanes, tornadoes, and severe storms.

2. *Tropical Rainfall Measuring Mission (TRMM) Satellite (NASA/JAXA):* TRMM is instrumental in monitoring and studying tropical rainfall, helping to understand and predict the behavior of tropical storms and hurricanes.

3. *Suomi NPP (National Polar-orbiting Partnership) Satellite (NASA/NOAA):* Suomi NPP contributes to monitoring and predicting climatic hazards by observing atmospheric conditions, sea surface temperatures, and land surface properties.

4. *COSMIC-2 (Constellation Observing System for Meteorology, Ionosphere, and Climate) Satellites (NASA/NOAA):* A constellation of satellites designed to improve weather forecasts and monitor atmospheric conditions, including tropical storms and cyclones.

5. *Sentinel-1 and Sentinel-2 (European Space Agency):* Part of the Copernicus program, Sentinel-1 employs synthetic aperture radar to monitor ground deformation, helping to detect and assess natural hazards such as earthquakes and landslides through high-resolution imagery.

6. *ALOS-2 (Japan Aerospace Exploration Agency):* Equipped with L-band synthetic aperture radar, ALOS-2 is used for disaster monitoring, including the observation of land deformation caused by earthquakes and volcanic activity.

7. *RADARSAT Constellation Mission (Canadian Space Agency):* A constellation of three satellites designed for Earth observation, RADARSAT monitors environmental changes and contributes to disaster management, including the monitoring of floods, earthquakes, and oil spills.

8. *DSCOVR (Deep Space Climate Observatory):* While primarily a solar observatory, DSCOVR also monitors Earth's climate and provides early warnings for solar storms, which can impact communication and power systems on Earth.

9. *Aqua Satellite (NASA):* Aqua carries instruments that contribute to the observation of climatic hazards, including those related to hurricanes and tropical storms, by monitoring sea surface temperatures and atmospheric conditions.

10. *SWOT (Surface Water and Ocean Topography) Satellite (NASA/ESA/CNES):* An upcoming satellite mission designed to observe changes in the world's oceans, lakes, and rivers. It will contribute to monitoring and understanding hazards such as floods and sea-level rise.

11. *MODIS (Moderate Resolution Imaging Spectroradiometer) on Terra and Aqua Satellites (NASA):* MODIS instruments on NASA's Terra and Aqua satellites provide data on land surface temperature, vegetation health, and changes in land cover that critical for monitoring the impact of drought on ecosystems.

12. *Landsat Satellites (NASA/USGS):* The Landsat program has been providing multispectral imagery for decades, enabling the monitoring of land cover changes, vegetation health, and drought impacts over time.

13. *TerraSAR-X (German Aerospace Center – DLR):* Terra SAR-X is a German radar satellite that can be used to monitor soil moisture and land deformation, providing valuable information for understanding drought conditions.

3.3 Importance of satellites for monitoring climate change hazards and impacts

Satellite remote sensing has provided major advances in understanding the climate system and its changes, by quantifying processes and spatio-temporal conditions of the atmosphere, land and oceans [22]. Satellites play a crucial role in mapping and monitoring climate change-related hazards and impacts. Remote sensing and modeling techniques can be used for rapid assessment of potentially endangered areas, such as high-mountain regions, where satellite remote sensing is a highly appropriate tool for monitoring purposes [28]. Additionally, satellites can be used to identify potential hazard sources and detect critical areas, which can be used to delineate endangered areas and inform land use planning, prevention, and mitigation measures [28]. Satellites can also be used to monitor and evaluate indicators for climate change-related health impacts, risks, adaptation, and resilience, providing valuable data for health professionals to understand, track, and manage climate change-related health risks [15]. Finally, satellites can be used to enhance early warning systems and ensure effective climate risk resilience development, particularly in vulnerable populations residing in rural communities [4].

4. Integration of space and ground-based data

4.1 The importance of integrated space and ground-based earth observations

Generally, the integration of space and ground-based data for assessing hazards and risk involves combining various types of data from different sources to improve the accuracy and efficiency of risk management strategies [29]. This approach has helped in better understanding to the various risks and have been increasingly used to support, providing valuable data and information products for domestic and international partners and stakeholders [30].

As a result, the United Nations Global Assessment Report 2022 Concept Note highlighted the need for accurate, timely, and relevant data and statistics for decision-makers, and consequently the Earth Observation Satellites (CEOS) Work Programme has been working on developing integrated geospatial and statistical methods to support risk-informed decision-making [31]. This showcasing the future-look of this approach that gives a comprehensive understanding of climate change's hazards and disaster risks by combining the strengths of both methods for reducing risk and improving resilience and ultimately contributing to more sustainable and resilient communities and ecosystems [32, 33].

4.2 Examples on the integration of space and ground-based earth observations

Combining space (SEOs)and ground-based Earth observations (GEOs) methods, will give a more comprehensive understanding of the complex interactions between natural and human-induced climate change's hazards and risk [31]. The following examples showing how the integration helps in assessing, managing and monitoring climate change's hazards and disaster risks:

1. *Wildland fire management*: The wildland fire hazard team identified seven major requirements for SEOs in fire management, including fuel mapping, risk assessment, detection, monitoring, mapping, burned area recovery, and smoke management [34]. The GEOs can provide detailed information on local-scale, such as fuel characteristics, while SEOs can provide large-scale information on fire extent and smoke plumes [35].

2. *Flood risk management in coastal zones*: Here, the GEOs offer ground's detailed information on local-scale, such as topography and land use, while the SEOs provide large-scale information on sea level rise and storm surge [9].

3. *Vulnerability and resilience assessment*: The integration helps in the assessment of communities and ecosystems vulnerability and resilience. GEOs offer detailed local information on the infrastructure and social systems, while SEOs provide large-scale data on land cover changes and atmospheric conditions [10].

4. *Erosion monitoring and assessment*: Merging the two methods help in monitoring and assessing erosion caused by wind and water, where GEOs offer detailed info on local-scale, such as soil characteristics, and SEOs provide large-scale data on land cover changes and precipitation patterns [11].

5. *Landslide monitoring*: GEOs offer detailed information on local-scale of landslide hazards, such as soil characteristics and topography, while SEOs provide large-scale evidence on land cover changes and precipitation patterns [36].

6. *Avalanche monitoring*: Avalanche hazards is exacerbated by climate change. GEOs detailed information on local-scale of the avalanche hazard, such as snowpack characteristics and topography, while SEOs provide large-scale information on snow cover changes and weather patterns [36].

7. *Drought monitoring*: GEOs offer detailed info on local-scale of drought hazards, such as soil moisture and vegetation health, while SEOs provide large-scale info on land cover changes and precipitation patterns [36].

8. *Air quality monitoring*: GEOs offer detailed information on the local-scale of air pollution, such as emissions sources and meteorological conditions, while SEOs provide large-scale data on atmospheric composition and transport [36].

9. *Glacier monitoring*: GEOs offer detailed info on local-scale assess glacier-based hazards, such as glacier mass balance and ice flow, while SEOs provide large-scale information on glacier extent and changes over time [36].

10. *Heatwave monitoring*: Heatwave hazards are becoming more frequent and severe due to climate change. GEOs provide detailed information on local-scale phenomena, such as urban heat islands and heat stress on vulnerable populations, while SEOs depicts the large-scale data on land surface temperature and atmospheric conditions [36].

11. *Tsunami monitoring*: Tsunami hazards are closely related to climate change. GEOs offer detailed info on its local-scale, such as seismic activity and coastal topography, while SEOs provide large-scale data on ocean surface changes and wave propagation [36].

12. *Tornado monitoring*: Tornado hazards, which are becoming more frequent and severe due to climate change. GEOs offer detailed information their local-scale, such as atmospheric instability and wind shear, while SEOs provide large-scale information on storm development and mesoscale features [36].

5. Climate change resilience through earth observations

Developing resilience for climate change's hazards and disaster risks involves assessing the hazards and vulnerabilities of communities and ecosystems, identifying areas and populations at higher risk, and developing effective strategies for reducing exposure, lessening vulnerability, and improving preparedness and early warning systems [4]. For instance, the "Reaching Resilience: Handbook Resilience 2.0" provided insights and ideas on how to design well integrated, step-by-step actions and strategies to foster resilience at the local level [37]. The following approaches are some ways to develop resilience for climate change's hazards and disaster risks:

1. *Improved hazard assessment*: SEOs such as satellite imagery are used to assess the extent of flooding and the impact of seismic events, providing essential information for disaster risk reduction [19].

2. *Enhanced early warning systems*: Space and ground-based Earth observations (SAGEOs) contributed to the development of early warning systems for climate-related hazards. Satellite data helped to monitor changes in sea surface temperature, enabling the early detection of conditions conducive to the formation of tropical cyclones, thus supporting timely and effective disaster preparedness and response [33].

3. *Support for risk-informed decision-making*: The integration of SAGEOs supports risk-informed decision-making by providing essential information for vulnerability and exposure mapping. For example, these observations help in identifying areas at risk from various hazards, thus informing adaptation planning and the development of climate-resilient strategies [19].

4. *Long-term climate data records*: Several SEOs provide long-term climate data records, essential for monitoring and quantifying slow changes in the climate system. These records are critical for understanding the evolving nature of climate-related hazards, such as changes in precipitation patterns, sea level rise, and the frequency of extreme weather events [33].

6. Hazards valuation and risk assessment

6.1 Quantifying climate change hazards and risks using SEOs

Space-based Earth observations, such as satellite imagery, enable the monitoring of various hazards, providing essential information for disaster risk reduction and vulnerability mapping which include:

1. *Risk assessment*: SEOs can help in assessing the risk of climate change hazards, such as floods, fires, and earthquakes, by providing large-scale information on land cover changes and atmospheric conditions [6].

2. *Detection and monitoring*: Satellite data can be used to detect and monitor climate change hazards, such as sea level rise, storm surges, and land cover changes, in real-time or near-real-time [34].

3. *Mapping fire fuel and burned area recovery*: SEOs is effortlessly used to map fuel types and features and assess the recovery of burned areas [34].

4. *Smoke management*: Satellite data can be used to monitor and assess smoke plumes from wildfires, providing valuable info for air quality management and public health [34].

5. *Climate change scenarios*: Through these observations we can develop climate change scenarios, which can help inform risk management procedures and support the development of effective strategies to enhance resilience and reduce risks [38].

6. *Mapping flood vulnerability index*: Through the multi-temporal satellite data flood vulnerability index could be developed to demonstrate the potential risk which supports flood risk management [32].

6.2 Evaluating climate change hazards and risks using GEOs

Evaluating climate change hazards and risks using GEOs involves two major approaches that rely on whether and giving more emphasis on the biophysical or the socio-economic aspect of climate-related risk as follows.

1. *Natural hazards-based approach*: This approach begins by characterizing the climate hazard and can be expressed as the probability of climate hazard multiplied by vulnerability. It focuses on the biophysical aspect of climate-related risk and is useful when the link between hazards and vulnerability is clear [39].

2. *Vulnerability-based approach*: This approach sets criteria based on socio-economic or biophysical outcomes and determines how likely these criteria are to be met or exceeded. It focuses on the socio-economic aspect of climate-related risk and is useful when the link between hazards and vulnerability is unclear or when the emphasis is on socio-economic outcomes [39].

In addition to the two above-mentioned approaches, there are other methods for quantifying the risks of climate change using GEOs data, such as:

- *Data-driven modeling*: This method uses machine learning algorithms to analyze large datasets and identify patterns and relationships between climate hazards, impacts, and risks [40].

- *Uncertainty quantification*: This method aims to quantify the uncertainty associated with climate change risk assessments, including future vulnerability and exposure to changing climate hazards [41].

- *Geospatial big data*: This method leverages geospatial data to better understand and quantify the risks of climate change, particularly in the context of extreme events [40].

7. Early warning systems and decision support

Early warning systems (EWS) are key elements of climate change adaptation and disaster risk reduction, and aim to avoid or reduce the damages caused by hazards. To be effective, EWS need to actively involve the people and communities at risk from a range of hazards, facilitate public education and awareness of risks, disseminate messages and warnings efficiently and ensure that there is a constant state of preparedness and that early action is enabled [42]. The significance of an effective EWS lies in the recognition of its benefits by local people.

Therefore, EWS for climate-related risks must rely on a sound scientific and technical basis such as SAGEOs data and focus on people or sectors mostly exposed to risk. This implies the adoption of a system approach incorporating all relevant risk factors, whether arising from the climate-hazards or social vulnerabilities, and from short-term or long-term processes. EWS include detection, analysis, prediction, and then warning dissemination followed by response decision-making and implementation.

Such systems are in place, in many parts of the world, to monitor, forecast, and warn people about the hazard such as tropical cyclones, floods, storms, tsunami, avalanches, tornadoes, severe thunderstorms, volcanic eruptions, extreme heat and cold, forest fires, drought, etc. Effective EWS consist of four interacting elements namely:

1. Risk knowledge,

2. Monitoring and warning services,

3. Dissemination and communication and

4. Response capability.

7.1 The role of early warning systems in decision-making

EWS play a crucial role in supporting decision-making to face climate change's hazards and disasters risk through SAGEOs. The 2021 state of climate services report highlighted the importance of EWS in reducing the impacts of climate change on human lives, ecosystems, and economies. The report made-out EWS as a top priority for the majority of countries, including 88% of least developed countries and small island states, that submitted their Nationally Determined Contributions (NDCs) to UNFCCC [43]. For instance, the Climate Risk and Early Warning Systems (CREWS) international initiative evolved to enhance early warning systems to assist vulnerable countries with climate change adaptation, and hence the CREWS-PNG project, came to develop an improved drought monitoring and early warning system to enable better strategic decision-making for agriculture, water management, health, and other climate-sensitive sectors [44].

Second example is the wildland fire hazard team produced a report, developed under the auspices of the Disaster Management Support Group (DMSG) of the G-7 Committee on Earth Observation Satellites, that reviewed potential requirements for SEOs in fire management. The team identified seven major requirements that contributed to the improvement of fire management programs [34].

These examples established the importance of EWS in supporting decision-making to face climate change's hazards and disasters risk through SAGEOs. They practically highlighted the need for effective communication, community participation, and the use of decision support systems to ensure timely and appropriate responses to climate change hazards.

7.2 Developing effective EWS for informed decision-making

Developing effective EWS is vital for informing timely and appropriate decisions to reduce the impact of disasters and emergencies through the following steps:

1. *Define warning objectives*: Clearly define the objectives of the EWS by determining the types of hazards, the target audience, and the desired outcomes, such as reducing casualties or minimizing property damage.

2. *Stakeholder engagement*: Involve key stakeholders, including government agencies, local communities, NGOs, and experts, in the development process, to ensure a collaborative approach that incorporates diverse perspectives.

3. *Technology and infrastructure*: Invest in suitable technology and infrastructure to support the EWS by including weather monitoring equipment, communication networks, and information dissemination tools.

4. *Data collection and monitoring*: Establishing a robust data collection and monitoring system using meteorological and geophysical data, satellite imagery, and real-time monitoring to track hazards and their impacts.

5. *Risk assessment*: Through conducting a comprehensive risk assessment to identify potential hazards and vulnerabilities in the target area. It helps to understand the historical context of disasters, assess the local environment, and consider the population's susceptibility and exposure.

6. *Risk communication*: Develop clear and effective communication strategies and tailoring messages to the needs of different audiences, considering language, cultural nuances, and literacy levels. Use multiple communication channels, including radio, TV, SMS, and social media.

7. *Warning criteria and triggers*: Define specific criteria and triggers that will activate the early warning system. This may involve establishing threshold values for certain parameters (e.g., river levels, wind speeds) that, when exceeded, trigger the issuance of warnings.

8. *Warning message formats*: Standardize the format of warning messages to ensure clarity and consistency. Include essential information such as the type of hazard, location, expected impacts, and recommended actions.

9. *Community engagement and training*: Conducting community outreach and training programs to raise awareness about the EWS and educate them on the meaning of different alerts and train them on appropriate response actions.

10. *Continuous evaluation and improvement*: Regularly evaluate the performance of the early warning system and seek feedback from stakeholders. This information will contribute in the improvements, update technologies, and adapt to changing risk profiles.

Through these steps, communities and authorities can establish effective EWS that enhance preparedness, response, and resilience in the face of natural disasters and emergencies. Thus, the integration of SAGEOs supports risk-informed decision-making by providing essential information for vulnerability and exposure mapping, informing adaptation planning, and the development of climate-resilient strategies.

8. Community engagement and citizen science

Community engagement and citizen science play a vital role in understanding and mitigating hazards and disaster risks. By involving local communities in data collection, risk assessment, and decision-making processes, citizen science initiatives can provide valuable insights and enhance the effectiveness of EWS [45]. These approaches not only contribute to more accurate risk assessments but also empower communities to take informed actions to reduce their vulnerability to disasters. Through collaborative research and data sharing, citizen science efforts enable the translation of research findings into public health actions that benefit the community, ultimately leading to more resilient and prepared communities.

8.1 Involving communities for enhanced disaster preparedness

The common approaches for engaging the community and citizen science for hazards valuation and disaster risk management in the era of climate change include the following:

1. *Community-centric approaches*: Emphasize the importance of community-driven goals and approaches, which tend to be more effective in co-creating a community vision of resilience, specifically as it relates to natural hazards and climate change. This involves engaging youth, artists, municipal officials, and community members in a variety of activities, including surveys, story-gathering, photovoice exhibits, forums, artist-created murals, and ripple effect mapping to increase knowledge of weather and climate, risks from local hazards, and strategies for mitigation [45].

2. *Bottom-up engagement*: This involves exploring pathways to improve disaster risk reduction through community participation, particularly in vulnerable communities affected by natural hazards and climate change. By involving local communities in the development and implementation of EWS, the approach will minimize the impacts of hazards on these communities [46].

3. *Integration of science and art*: Incorporate a mixture of art, science, and community engagement to provide a range of opportunities for personal connection and learning. This approach helps participants with meaning-making about local hazards and assets, allowing for a sense of familiarity and interconnectedness, and creating a shared vision of resilience [45].

4. *Capacity strengthening*: Focus on strengthening the capacity of community managed disaster risk reduction committees and communities through training and assessment. This involves conducting Climate Vulnerability and Capacity Analysis (CVCA) training to achieve the links between development and climate-related hazards and disasters by engaging the communities on climate vulnerability and capacity analysis [46].

8.2 Necessity of engaging the community in hazards valuation

The importance of engaging the community and citizen science in hazards valuation and disaster risk management in the time of climate change is matter of concern. The following key insights will demonstrate this necessity:

- *Community resilience*: Engaging the community in co-creating a vision of resilience, particularly in the context of natural hazards and climate change, is essential. This involvement can lead to increased knowledge of weather and climate, local hazards, and strategies for mitigation, ultimately fostering a sense of familiarity and interconnectedness within the community [45].

- *Vulnerable communities*: In the face of climate change and natural hazards, vulnerable communities are disproportionately affected. Engaging them through EWS and risk management approaches is crucial for minimizing the impacts of disasters and building resilience [46].

- *Socioeconomic parameters*: Hazards and vulnerability are not only associated with natural phenomena but also with socioeconomic parameters. Engaging the community can help in better understanding the socioeconomic aspects of vulnerability, which is often overlooked in risk assessments [11].

Consequently, engaging the community and citizen science is important for fostering resilience, minimizing the impacts of disasters on vulnerable communities, understanding the socioeconomic aspects of vulnerability, and addressing the global recognition of the importance of EWS in disaster risk management.

9. Modeling the future of climate change and hazards

Modeling the future of climate change and hazards using SAGEOs involves various techniques and approaches (**Table 1**). These techniques help to understand, predict, and mitigate the impacts of climate change and hazards on human populations and the environment. Some key aspects of modeling the future of climate change and hazards include:

- *Integration of SAGEOs*: The Integrated Global Observing Strategy (IGOS) is an international initiative that calls for an integrated use of observations from a multitude of platforms, including ground-based, balloon, aircraft, and satellite observations, as well as numerical modeling efforts [51].

- *Big earth data*: Big Earth data, which includes all data related to Earth science, provides technical computing resolution on Earth-related data and enables rapid and accurate monitoring of the Earth system. This data can be used to model the Earth's interior, surface, atmosphere, and socioeconomic data [52].

- *Community engagement and citizen science*: Involving different techniques for data analysis and system modeling is more appropriate for practical decision-making than a purely reductionist approach [53].

- *Early warning systems*: SEOs can be used to improve EWS for natural disasters, such as wildfires through fuel mapping, risk assessment, detection, monitoring, mapping, burned area recovery, and smoke management [34].

- *Atmospheric greenhouse gas monitoring*: SEOs can provide info over regional areas and with relatively low cost, adding synoptic and geospatial extent to

Modeling	Description	Data needed	References
Flood Risk modeling	Modeling floods to identify areas at risk of flooding	Precipitation, soil moisture, river discharge	[32, 47, 48]
Wildfire Risk modeling	Modeling wildfires to identify areas at risk of wildfires	Temperature, humidity, wind speed, vegetation moisture content	[14, 35, 49]
Drought Risk modeling	Modeling droughts to identify areas at risk of droughts	Precipitation, soil moisture, vegetation health	[13, 33]
Landslide risk modeling	Modeling landslides to identify areas at risk of landslides	Rainfall, soil moisture, topography	[19, 36]
Extreme weather risk modeling	Predicting extreme weather events to develop models that identify areas at risk of these events	Temperature, humidity, wind speed, precipitation	[50]
Seismic hazard modeling	Modeling and predicting seismic hazards to develop models that identify areas at risk of earthquakes	Seismic activity, ground deformation, geological structure	[48]
Tropical cyclone risk modeling	Predicting tropical cyclones to develop models that identify areas at risk of cyclones	Sea surface temperature, wind patterns, atmospheric pressure	[49]
Carbon monitoring	Understanding the evolution of carbon, its sources and sinks, and its impact on climate change to develop models that enable proactive measures to mitigate risks	Carbon dioxide concentrations, land use changes, ocean acidification	[33]

Table 1.
Showing how space and ground-based earth observations can be used to model the future of climate change's hazards and disasters.

ground-based measurements. This is particularly important for monitoring atmospheric greenhouse gases, such as carbon dioxide (CO2) and methane (CH4), which play a major role in the Earth's climate system [54].

10. Conclusion

In the era of climate change, assessing hazards and disaster risk requires the integration of space-based and ground-based Earth observations through two main approaches: natural hazards-based and vulnerability-based approaches. Both approaches can be used separately or together, depending on the specific context and available data. Space-based Earth observations, such as satellite-based remote sensing applications, can support disaster risk management by providing information on past events, physical processes associated with hazards, and ultimately benefiting populations at risk from disasters. Ground-based methods, such as event tree analysis, can be used to develop a common methodology for risk assessment and mapping of climate change-related hazards. By integrating space and ground-based Earth observations, it becomes possible to quantify hazards, assess the risks of climate change, and develop effective strategies for disaster risk reduction and climate resilience making them a cornerstone of efforts towards a sustainable and resilient future.

Acknowledgements

Authors are appreciating the help of African Center of Excellence on Climate Change, Biodiversity and Sustainable Agriculture (CEA-CCBAD), at University Felix Houphouet-Boigny (UFHB), Côte d'Ivoire and the support of the Partnership for Skills in Applied Sciences, Engineering and Technology.

Conflict of interest

The authors declare no conflict of interest.

Author details

Nasradeen A.H. Gadallah[1,2]* and Ahmed A.H. Siddig[1,3]

1 Faculty of Forestry, Department of Forest Conservation and Protection, University of Khartoum, Khartoum, Sudan

2 African Center of Excellence on Climate Change, Biodiversity and Sustainable Agriculture, University Felix Houphouet-Boigny, Abidjan, Côte d'Ivoire

3 Harvard Forest, Harvard University, Petersham, Massachusetts, USA

*Address all correspondence to: naouofk@gmail.com

References

[1] Bruen M, Dzakpasu M. A Review of Climate Change-Related Hazards and Natural Disaster Vulnerabilities and of Agencies Involved in Warning and Disaster Management. Dublin, Ireland: Environmental Protection Agency; 2018. pp. 1-95

[2] Lavell A, Oppenheimer M, Diop C, Hess J, Lempert R, Li J, et al. Climate change: New dimensions in disaster risk, exposure, vulnerability, and resilience. In: Field CB, Barros V, Stocker TF, Dahe Q, editors. Managing the Risks of Extreme Events and Disasters to Advance Climate Change Adaptation. 1st ed. Cambridge, UK: Cambridge University Press; 2012. pp. 25-64

[3] Pulhin JM, Inoue M, Shaw R, editors. Climate Change, Disaster Risks, and Human Security: Asian Experience and Perspectives. Singapore: Springer; 2021. 133 p

[4] Agbehadji IE, Schütte S, Masinde M, Botai J, Mabhaudhi T. Climate risks resilience development: A bibliometric analysis of climate-related early warning Systems in Southern Africa. Climate. 2023;**12**(1):3

[5] Onuoha DC, Ogbo OG, Amaechi M. The need for resilient infrastructure as an adaptive measure for climate change. BJES. 2022;**10**(4):17-27

[6] John P, Kryspin-Watson J, Nieuwejaar S. Disaster Risk Management and Climate Change Adaptation in Europe and Central Asia. Washington, USA: World Bank; 2010

[7] Solecki W, Leichenko R, O'Brien K. Climate change adaptation strategies and disaster risk reduction in cities: Connections, contentions, and synergies. Current Opinion in Environmental Sustainability. 2011;**3**(3):135-141

[8] Jones RL, Kharb A, Tubeuf S. The untold story of missing data in disaster research: A systematic review of the empirical literature utilising the emergency events database (EM-DAT). Environmental Research Letters. 2023;**18**(10):103006

[9] Cienfuegos R. Flood risk from geophysical and hydroclimatic hazards: An essential integration for disaster risk management and climate change adaptation in the coastal zone. Natural Hazards. 2023;**119**(2):1113-1115

[10] Fekete A. Disaster risk, climate change, and urbanization as research topics in Western Asia: A bibliometric literature analysis. Climate. 2023;**11**(6):131

[11] Gilard O. Hazards, vulnerability and risk. In: Climate Change and Agriculture Worldwide. Dordrecht, Netherland: Springer; 2016. pp. 19-29

[12] Shi P, Shi P. Hazards, disasters, and risks. In: Disaster Risk Science. China, Beijing: Beijing Normal University Press, and Springer Nature Singapore Pte Ltd; 2019. pp. 1-48

[13] Ghozat A, Sharafati A, Hosseini SA. Satellite-based monitoring of meteorological drought over different regions of Iran: Application of the CHIRPS precipitation product. Environmental Science and Pollution Research International. 2022;**29**(24):36115-36132

[14] Abrha H, Dodiomon S, Ongoma V, Hagos H, Birhane E. Spatio-temporal prediction of climate and wildfire in Hugumbrda Grat-Kahsu forest,

Tigray: Priority for early warning. Geomatics, Natural Hazards and Risk. 2023;**14**(1):2250517

[15] Ebi KL, Boyer C, Bowen KJ, Frumkin H, Hess J. Monitoring and evaluation indicators for climate change-related health impacts, risks, adaptation, and resilience. International Journal of Environmental Research and Public Health. 2018;**15**(9):1943

[16] Landrigan PJ, Britt M, Fisher S, Holmes A, Kumar M, Mu J, et al. Assessing the human health benefits of climate mitigation, pollution prevention, and biodiversity preservation. Annals of Global Health. 2024;**90**(1):1

[17] Uitto J. Evaluating Environment in International Development. Milton Park, Abingdon, Oxon; New York, NY: Routledge; 2021. 346 p

[18] Hochrainer-Stigler S, Velev S, Laurien F, Campbell K, Czajkowski J, Keating A, et al. Differences in the dynamics of community disaster resilience across the globe. Scientific Reports. 2021;**11**(1):17625

[19] Le Cozannet G, Kervyn M, Russo S, Ifejika Speranza C, Ferrier P, Foumelis M, et al. Space-based earth observations for disaster risk management. Surveys in Geophysics. 2020;**41**(6):1209-1235

[20] National Research Council. Observing Weather and Climate from the Ground Up: A Nationwide Network of Networks. Washington, DC 20433 USA: National Academies Press; 2009. 250 p

[21] Aschbacher J, Milagro-Pérez MP. The European earth monitoring (GMES) programme: Status and perspectives. Remote Sensing of Environment. 2012;**120**:3-8

[22] Yang J, Gong P, Fu R, Zhang M, Chen J, Liang S, et al. The role of satellite remote sensing in climate change studies. Nature climate change. 2013;**3**(10):875-883

[23] Hegglin MI, Bastos A, Bovensmann H, Buchwitz M, Fawcett D, Ghent D, et al. Space-based earth observation in support of the UNFCCC Paris agreement. Frontiers in Environmental Science. 2022;**10**:941490

[24] Duvat VK, Magnan AK, Wise RM, Hay JE, Fazey I, Hinkel J, et al. Trajectories of exposure and vulnerability of small islands to climate change. Wiley Interdisciplinary Reviews: Climate Change. 2017;**8**(6):478

[25] Young OR, Onoda M. Satellite earth observations in environmental problem-solving. In: Satellite Earth Observations and their Impact on Society and Policy. Singapore: Springer; 2017. pp. 3-27

[26] Burke M, Driscoll A, Lobell DB, Ermon S. Using satellite imagery to understand and promote sustainable development. Science. 2021;**371**(6535):8628

[27] Nesterenko S, Shariy G, Shchepak V, Tkachenko I, Trifonova A. Geomatic monitoring of environmental hazards in technogenic-loaded territories. Land management, cadastre and land monitoring. 2023;**2023**(2):1-24

[28] Huggel C. Impacts from climate change related hazards in high-mountain areas: A review of assessment techniques. In: First International Conference on the Impact of Climate Change on High-Mountain Systems, 21-23 November 2005, Bogota, Colombia. Russian Academy of Sciences; 2005. pp. 109-121

[29] Dimitrova M, Snair M. Classifying disaster risk reduction strategies: Conceptualizing and testing a novel integrated approach. Globalization and Health. 2024;**20**(1):1-15

[30] Bell J. NASA Earth Science Disasters Program. Huntsville, USA: University of Alabama in; 2017

[31] Borges DE, Ramage S, Green D, Justice C, Nakalembe C, Whitcraft A, et al. Earth observations into action: The systemic integration of earth observation applications into national risk reduction decision structures. Disaster Prevention and Management: An International Journal. 2023;**32**(1):163-185

[32] Cian F, Giupponi C, Marconcini M. Integration of earth observation and census data for mapping a multi-temporal flood vulnerability index: A case study on Northeast Italy. Natural Hazards. 2021;**106**(3):2163-2184

[33] Gaetani F, Petiteville I, Pisano F, Rudari R, St PL. A synergy framework for the integration of earth observation technologies into disaster risk reduction. In: EGU General Assembly Conference Abstracts. European Geosciences Union (EGU) [Online]; 2015. p. 15780

[34] Lynham TJ, Dull CW, Singh A. Requirements for space-based observations in fire management: A report by the wildland fire Hazard team, committee on earth observation satellites (CEOS) disaster management support group (DMSG). IEEE International Geoscience and Remote Sensing Symposium. 2002;**2**:762-764

[35] Nakata M, Sano I, Mukai S, Kokhanovsky A. Characterization of wildfire smoke over complex terrain using satellite observations, ground-based observations, and meteorological models. Remote Sensing. 2022;**14**(10):2344

[36] Singh RP, Bartlett D, editors. Natural Hazards: Earthquakes, Volcanoes, and Landslides. 1st ed. Boca Raton, Florida: Taylor & Francis Group, the academic division of T&F Informa plc; 2018. 78 p

[37] Heijmans E. Reaching Resilience: Handbook Resilience 2.0 for Aid Practitioners and Policymakers in Disaster Risk Reduction, Climate Change Adaptation and Poverty Reduction. Netherland: CARE Nederland, Groupe URD and Wageningen University; 2013. 127 p

[38] Zeidler A, Dobesberger P. Considering climate change in natural Hazard risk Management in Energy Systems a practical approach. In: Extended Abstracts. Innsbruck, Austria: Federal Research and Training Centre for Forests, Natural Hazards, and Landscape (BFW); 2012. pp. 446-447

[39] Jones R, Boer R, Magezi S, Mearns L. Technical paper 4: Assessing current climate risks. Adaptation policy frameworks for climate change: developing strategies, policies and measures. 2004;**4**:93-117

[40] Jones A, Kuehnert J, Fraccaro P, Meuriot O, Ishikawa T, Edwards B, et al. AI for climate impacts: Applications in flood risk. npj Climate and Atmospheric Science. 2023;**6**(1):1-8

[41] Harrington LJ, Schleussner CF, Otto FEL. Quantifying uncertainty in aggregated climate change risk assessments. Nature Communications. 2021;**12**(1):7140

[42] EEA EEA. Late Lessons from Early Warnings: Science, Precaution, Innovation: Summary. Copenhagen, Denmark: European Environment Agency; 2013. 58 p

[43] Grasso V. State of climate services report: Risk information and early warning systems. In: EGU General Assembly Conference Abstracts. Vol. 2021. European Geosciences Union (EGU) [Online]; 2020. pp. 21-74

[44] Kuleshov Y, Inape K, Watkins AB, Bear-Crozier A, Chua ZW, Xie P, et al. Climate risk and early warning systems (CREWS) for Papua New Guinea. In: Ondrasek G, editor. Drought–Detection and Solutions. London, UK: IntechOpen; 2020. pp. 147-168

[45] Semmens KA, Carr RH, Maxfield K, Sickler J. CREATE resilience through science, art, and community engagement. Community Science. 2023;**2**(3):1-17

[46] Senaratna N, Baudoin MA, Oluoko-Odingo AA, Ajuang L, Wepukhulu DW, Mwadali AS. In: Singh A, Zommers Z, editors. Natural Hazards and Climate Change in Kenya: Minimizing the Impacts on Vulnerable Communities through Early Warning Systems. Dordrecht: Springer Netherlands; 2014. pp. 355-375

[47] Hirabayashi Y, Tanoue M, Sasaki O, Zhou X, Yamazaki D. Global exposure to flooding from the new CMIP6 climate model projections. Scientific Reports. 2021;**11**(1):3740

[48] Weatherhead EC, Wielicki BA, Ramaswamy V, Abbott M, Ackerman TP, Atlas R, et al. Designing the climate observing system of the future. Earth's Future. 2018;**6**(1):80-102

[49] Colbert A. Extreme Weather and Climate Change. Climate Change: Vital Signs of the Planet. Available from: https://climate.nasa.gov/extreme-weather [Accessed: January 23, 2024]

[50] Jones R, Mearns L. Assessing future climate risks. In: Adaptation Policy Frameworks for Climate Change: Developing Strategies, Policies and Measures. UK: Cambridge University Press; 2004. pp. 119-143

[51] Mazière M, Colin R, Muer DD, Zander R. The earth observing system: Synergy between ground-based and satellite observation, and model experiments. In: Space Scientific Research in Belgium. Belgium: Space Scientific Research; 2017. pp. 9-32

[52] Guo H. From digital earth to big earth data: Accelerating scientific discovery and supporting global sustainable development. International Journal of Digital Earth. 2019;**12**(1):1-1

[53] Hunt JCR, Aktas YD, Mahalov A, Moustaoui M, Salamanca F, Georgescu M. Climate change and growing megacities: Hazards and vulnerability. Proceedings of the Institution of Civil Engineers—Engineering Sustainability. 2018;**171**(6):314-326

[54] Sai S. Retrieval and Study of Columnar CO2 and CH4 Using Ground-Based FTIR Spectrometer [Master]. Hyderabad, India: International Institute of Information Technology; 2022

Chapter 11

Perspective Chapter: Engaging Indigenous Peoples – The SERVIR Amazonia Experience

Thomas Moore

Abstract

This chapter presents the experience of the SERVIR Amazonia program for the transfer of geospatial technology with efforts to include indigenous peoples and their organisations in its services. The overall program objectives are to contribute to the conservation of biological diversity and to the mitigation of or adaptation to climate change. Much of Amazonia where biological diversity is greatest and climate change threats are highest is indigenous peoples' territories; therefore, their inclusion in program services is critical to success. This analysis highlights the approach, advances, and challenges, of those efforts, including the ethical considerations and risks involved. Such efforts must include strong cultural sensitivity and communication skills. Among the conclusions are that the approach of working with and through the indigenous peoples' own representative organisations is highly appropriate, especially the most local ones.

Keywords: SERVIR Amazonia, geospatial technology, social inclusion, indigenous peoples, forest monitoring, cultural sensitivity

1. Introduction

This analysis provides a record of the experience of the SERVIR Amazonia with regard to the inclusion of indigenous peoples in the geospatial services that it provides under its mandate. It addresses the challenges involved, the approach adopted, and the successes and limitations of that experience. It highlights the need for cultural sensitivity and the use of appropriate technology and vocabulary. This chapter updates and expands the range of focus of a previous publication on this subject by the author [1].

2. The program

2.1 SERVIR Amazonia

SERVIR Amazonia is a joint iniative of the [United States] National Aeronautics and Space Agency (NASA) and the United States Agency for International Development (USAID). Its overall purpose is to contribute to the conservation of biological diversity and to the mitigation of and/or adaptation to climate change through the transfer of geospatial technology, developed by NASA and other institutions, as appropriate, to government and private sector entities active in the countries where it is active. Its slogan "from space to village" reflects that purpose. SERVIR Amazonia is articulated into the SERVIR Global network, which began in 2005 with the SERVIR Mesoamerica program, which has concluded. It makes available NASA-generated and other geographical information services (GIS) at no cost to underserved users in the regional networks. It currently includes two African programs, two Asian programs, and a planned Caribbean program in addition to SERVIR Amazonia.

The principal associates of SERVIR Amazonia are the International Centre for Tropical Agriculture (CIAT), the lead organisation, based in Cali, Colombia; the Association for the Conservation of the Amazon Basin (ACCA), in Lima, Peru; the Institute for the Certification of Forestry and Agricultural Management (IMAFLORA), in Piracicaba, São Paulo, Brazil; the Ecociencia Foundation (ECOCIENCIA), in Quito, Ecuador; and the Spatial Informatics Group (SIG) in Pleasanton, California, United States. It also includes initiatives in Guyana and in Suriname, coordinated by CIAT. And it sponsors an academic partner network, led by the University of Arkansas, in the United States.

SERVIR-Amazonia supports sustainable development throughout the Amazon region by strengthening the capacity of governments, the private sector, and other key stakeholders, including indigenous peoples' organisations, to integrate Earth observation and geospatial technologies into improved decision-making and to better incorporate the voices of women, indigenous peoples and their communities.

The first phase of SERVIR Amazonia began on December 21, 2018 and concluded on December 20, 2023. A follow-on effort began on December 21, 2023, and will continue for three years for the purpose of expanding the consolidation of geospatial information services in Brazil, Colombia. Ecuador, and Peru.

For more detail on the SERVIR Amazonia program, its team, services, data catalogue, and publications, see its webpage [2].

2.2 SERVIR's Amazonia strategy for the engagement of indigenous peoples

At its start-up workshop, held at CIAT headquarters near Cali, Colombia, March 4–8, 2019, SERVIR Amazonia adopted its Strategy for the Engagement of Indigenous Peoples [3]. This Strategy emphasises 1) collective empowerment of indigenous individuals to intelligently engage in decision-making processes; 2) leveraging and improving the existing capacity of the indigenous peoples' communities and organisations (IPO) and reinforcing their networks to connect existing capacity with regional needs by developing opportunity structures among them; and 3) normalising geospatial information and its uses across a broader set of individuals than typical expert users, including IPO leadership and technical staff.

It contemplates at a minimum:

- Training of SERVIR Amazonia and core partner staff on indigenous peoples' values, needs, and priorities, as these relate to the conservation of biological diversity within and around their territories, their perspectives on the development of their own and external economies and interests;

- Outreach to the Coordination of Indigenous Organisations in the Amazon Basin (COICA) and its member IPOs to inform them of opportunities in SERVIR Amazonia activities;

DOI: http://dx.doi.org/10.5772/intechopen.1004316

- Inclusion of the IPOs in mapping of areas of SERVIR Amazonia where the IPOs can make effective use of geospatial technology and information.

- Provision of information on available geospatial services and their utility for informing both their own decisions regarding land, forest, and water use, and detection of incursions into their territories by gold miners, timber pirates, drug traffickers, colonist invaders, and other external actors;

- Developing criteria for effectively connecting geospatial technology and services with the IPOs in each of the specific geographical areas of SERVIR Amazonia intervention;

- Provision of technical assistance to IPO representatives and GIS staff in the use of geospatial services in the selected geographic areas.

- Provision of technical assistance to appropriate technical instances of country governments in USAID's Regional Environmental Program for Amazonia (AREP) to allow them to incorporate information derived from IPO sources into their development planning in a manner that supports indigenous peoples' values, needs, and priorities in their planning and operations;

- Access to vehicles for dissemination of information derived from geospatial technology and tools that raise public awareness of the threats to indigenous territories and options for preventing or mitigating them in a transparent manner.

This Strategy was implemented during the five years of the first phase of SERVIR Amazonia and continues to apply to the follow-on initiative.

3. The indigenous peoples, territories, and organisations

Much of the available official information on South American indigenous populations is unreliable and reflects undercounts, particularly for those residing outside officially recognised indigenous territories or in urban areas. We have researched the topic over many years with input from a wide range of sources, including government reports, the IPOs themselves, international cooperation reports and others, and we compiled our own data in a 2021 analysis of Amazonian indigenous peoples reproduced in the chart below.

The numbers reported reflect self-identified peoples of Amazonian origin; they do not include indigenous peoples of highland Andean origin or other recent migrants, who merit separate attention and consideration. Indigenous peoples present in two or three countries are counted only once in the totals. The principal sources were the COICA member organisations, governments reports, data obtained from non-governmental organisations in the countries and the International Work Group for Indigenous Affairs (IWGIA) in Copenhagen, Denmark. The interpretations and final tally are the author's [4] (**Table 1**).

Amazonian indigenous peoples speak approximately 300 different languages within 25 established language families and include an additional 25 language isolates having no conclusively established relationship with other language families [5, 6].

Country	Peoples	Individuals	Sources	Dates
Bolivia	34	320,000	CIDOB, VMA, IME	2020
Brazil	256	950,000	COIAB, FUNAI, IBGE, ISA	2020
Colombia	26	52,000	OPIAC, DANE	2018, 2020
Ecuador	10	130,000	CONFENIAE, INEC	2018, 2020
French Guiana	6	13,000	FOAG, INSEE	2018, 2020
Guyana	9	82,000	APA, GBS Census	2012, 2019
Peru	51	550,000	AIDESEP, INEI	2017, 2020
Suriname	25	21,000	OIS, VIDS, ABS	2012, 2020
Venezuela	51	740,000	ORPIA, INE	2020
Totals	431	2,858,000	IWGIA by country	

Source abbreviations: ABS, Algemeen Bureau voor Statistiek, Suriname; AIDESEP, Asociación Indígena para el Desarrollo de la Selva Peruana; APA, Amerindian Peoples Association, Guyana; CIDOB, Confederación de Pueblos Indígenas del Oriente Boliviano; COIAB, Coordenação das Organizacões Indígenas da Amazônia Brasileira; CONFENIAE, Confederación de Nacionalidades Indígenas de la Amazonía Ecuatoriana; DANE, Departamento Administrativo Nacional de Estadísticas, Colombia; FOAG, Fédération des Organisations Amérindiens de Guyane, France; FUNAI, Fundação Nacional do Indio, Brasil; GBS, Guyana Bureau of Statistics; INE, Instituto Nacional de Estadística, Bolivia, Venezuela; INEC, Instituto Nacional de Estadística y Censos, Ecuador; INEI, Instituto Nacional de Estadística e Informática, Peru; INSEE, Institut Nacional de la Statistique et des Études Économiques, France; ISA, Instituto Socioambiental, Brazil; IWGIA, International Work Group for Indigenous Affairs, Denmark; OIS, Organisatie van Inheemse Volken in Suriname; OPIAC, Organización de Pueblos de la Amazonía Colombiana; ORPIA, Organización Regional de Pueblos Indígenas de Amazinas, Venezuela; VIDS, Vereniging van Inheemse Dorpshoofden, Suriname; VMA, Vice-Ministerio de Autonomías, Bolivia.

Table 1.
Amazonian indigenous peoples.

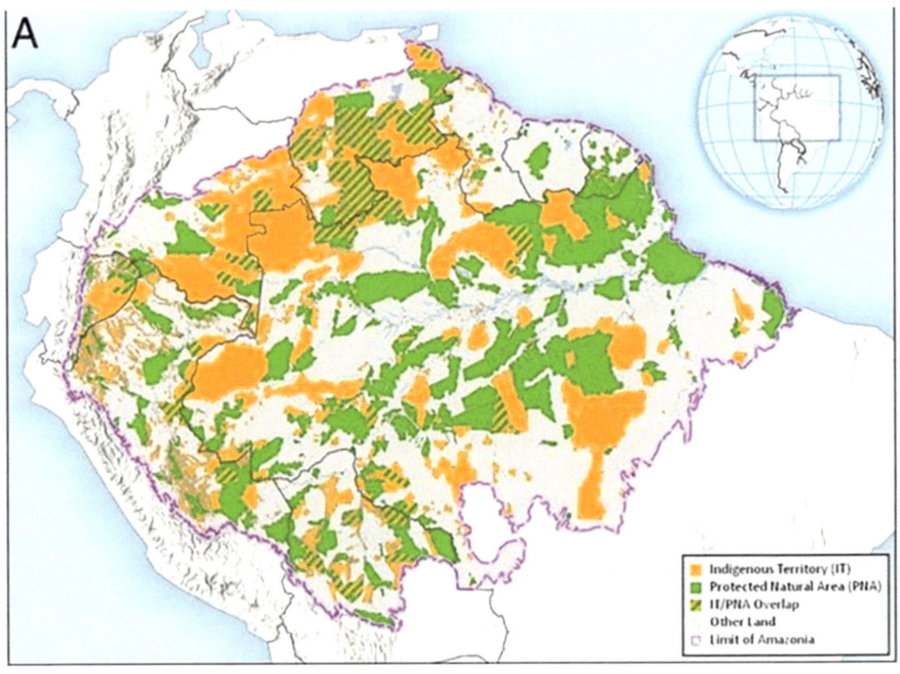

Figure 1.
Amazonian indigenous territories and protected natural areas.

Their traditionally conserved forests, along with those of natural protected areas, provide the greatest stores of diverse biological species and favourable opportunities for addressing climate change [7–9]. Moreover, they are effective buffers to deforestation throughout Amazonia [10–12]. The map below shows the locations of the officially recognised indigenous territories in the Amazonian countries (**Figure 1**) [8].

On an Amazon Region level, the indigenous peoples are organised in COICA, which has member organisations in each of the nine countries: CIBOB in Bolivia, COIAB in Brazil, OPIAC in Colombia, CONFENIAE in Ecuador, FOAG in French Guiana, APA in Guyana, AIDESEP in Peru, OIS in Suriname, and ORPIA in Venezuela, each of which has numerous affiliated more local federations that represent most of the local indigenous communities in each country. In Brazil, Colombia, Ecuador, and Peru there are also national-level organisations that include Andean and other non-Amazonian indigenous peoples, and there are other national-level organisations, the Confederación de Nacionalidades Amazónicas del Peru (CONAP) and VIDS in Suriname, that must also be taken into consideration, along with their affiliated organisations.

4. The principal experiences and successes

The first SERVIR Amazonia training experience was conducted in Guyana during 2019, where the University of Guyana priority service was for monitoring of mangroves along the northwestern Caribbean coast. APA provided one male indigenous leader and one male member of their technical team for the two training sessions in basic GIS technology which they found useful beyond the purposes of the mangroves monitoring. APA has one member community in the mangroves area, but most of their communities are located in the southern Rupununi Region. Efforts to develop a more complete service directly with APA and their communities did not come to fruition, because potential SERVIR Amazonia partner organisations working in Guyana were unable to obtain complementary funding for computers and infrastructure.

More detailed training was provided indigenous communities and their organisations in Peru, Brazil, Colombia, and Ecuador. In Peru, SERVIR Amazonia entered into a memorandum of understanding with the Madre de Dios Native Federation (FENAMAD), whereby ACCA provided three week-long training sessions on a range of GIS programs, including Google Earth, Global Forest Watch, Planet, Synthetic Aperture Radar (SAR) and Light Detection and Ranging (LIDAR) applications in rented facilities where appropriate computers were available. The participants demonstrated keen observation and quick learning skills that they will use in future initiatives to strengthen their ability to conserve and defend their collective territories.

Participants included members of FENAMAD's leadership team, technical staff and, particularly, volunteer community forest monitors. A total of 26 indigenous participants, completed this training with a focus on identifying and documenting illegal gold mining, logging, and drug trafficking by external actors present within their territories. Using one or more of these techniques and programs, community forest monitors locate the threats to the community territory with a global positioning system (GPS). Then they communicate that information to the FENAMAD technical team in Puerto Maldonado. Following analysis and approval by the FENAMAD governance Council, a FENAMAD lawyer presents this documentation to the appropriate agencies, such as Peruvian Government's environmental prosecutors (FEMA) for follow-on prosecution procedures or to Regional Government departments like Forestry and Wildlife, Energy and Mines, or Agriculture, for administrative action.

Sometimes, when appropriate, the press is called in, and the information is published on the FENAMAD website. Such situations must be weighed carefully and decided upon collectively following extensive internal discussion. There are obvious risks to accusing illegal actors of their crimes.

Similar, but less sophisticated, training was provided to CONFENIAE leaders in Ecuador by ECOCIENCIA and to Muinane Witoto communities in Colombia by Instituto Sinchi, under SERVIR Amazonia contracts. In all cases the indigenous participants defined their own training priorities and located them within their pre-determined development objectives, and the training responded to their needs as they explained them to the SERVIR Amazonia scientists, who thereby became more familiar with the problems the indigenous peoples were experiencing and helped them respond more effectively.

The most successful SERVIR Amazonia experience with the engagement of indigenous peoples was conducted by the University of Richmond, Virginia's, Applied Science Team in the Ucayali Region of Peru and Acre State in Brazil. It was led by geographers Stephanie Spera and David Salisbury. In those areas, more than 200 indigenous leaders and community members affiliated with the Association for the Integral Development of the Yurua-Yono-Sharakoiu Region (ACONADIYSH) in Ucayali, Peru, and the Organisation of Indigenous Peoples of the Juruá River (OPIRJ) in Acre, Brazil, learned to apply geodashboards in the monitoring of forest roads (MOCAF), and in ecosystem services and evapotranspiration. Meteorological stations were set up and used continuously by the indigenous communities, as well as the appropriate government agencies and non-profit organisations.

For these purposes, the University of Richmond team incorporated training support from ACCA and the Upper Amazon Conservancy (UAC) in Peru and the Acre Pro-Indian Comission in Brazil. That experience culminated in a conference at the University of Richmond, Virginia, in which seven indigenous leaders participated, followed by a fund-raising and informational trip to Washington, D. C., in November 2023.

5. The challenges

SERVIR Amazonia began by meeting with the leaders of COICA and its member national organisations, who were invited to participate in the initial program launch workshops in each of the countries where it worked. However, it soon became evident that some of the more local organisations affiliated with these organisations could be more effectively included in the program services in the areas of their presence, while keeping the national organisations and COICA informed of these activities.

COICA and the national-level indigenous organisations operate with modest organisational and financial resources. Moreover, SERVIR Amazonia lacks both the mandate and the budget to provide needed infrastructure and computers to the indigenous organisations, which depended upon other international cooperation funding for their operations and activities; so, they could not provide these resources themselves. SERVIR Amazonia is intended to be a network to foster the use of geospatial services, not a funding source for IPOs or non-profit organisations that work with them. Thus, funding to cover infrastructure and computers must be sought elsewhere. Training, however, is a legitimate SERVIR Amazonia contribution; so, most of SERVIR Amazonia's support to the IPOs was for training purposes.

However, it was possible to include some of the principal indigenous IPOs leaders and community members in services developed by other public and private organisations, as well as the more local indigenous federations in training programs developed around some of the principal services organised in response to the needs of other public and private organisations, and also to schedule small-scale services under memoranda of understanding with the indigenous federations themselves in Peru and Brazil.

The indigenous organisations and their communities prioritised territorial monitoring and mapping in order to more effectively defend their territories from outside incursions by gold miners, loggers, drug traffickers, land traffickers, traffickers in persons, and unwanted adventurers. So, their own volunteer forest monitors and their organisation leaders and technical staff were prioritised as beneficiaries. Other options for geospatial support sought by the IPOs included the monitoring of fire, floods, and other natural disasters.

Many of the indigenous participants lacked adequate and well-maintained personal computers and appropriate software, or else these were shared by families, so that they were not available when their children needed the computers in school or similar situations arose. Moreover, many of the local communities are located in remote rural areas, where they have no access to electricity or internet signals; so, most of the training activities had to be conducted in larger towns or cities in their regions where rented facilities were available, and the participants had to be brought in from their communities and provided with lodging and meals as well as travel costs. Also, most of the participants had very limited computer skills and lacked experience with online program options and GIS; so, the training had to include learning these skills, and this was done.

Ethical considerations must also be considered and assumed, particularly as regards data ownership, stewardship, and sharing. For example, in 2020, AIDESEP withdrew from discussion of a SERVIR Amazonia-supported grant proposal to integrate and consolidate their database on indigenous territories among their nine regional organisations throughout the Peruvian Amazon, because access to the database would have been public, and therefore available to their adversaries in land disputes or to illegal actors interested in the natural resources in their forested territories. Also, the application of geospatial technology brings with it culture change, which can be either harmful or helpful [13] and mapping, if not controlled by the indigenous peoples themselves, can inadvertently unite or divide kinship groups or other indigenous social organisations for official or private purposes, often without their knowledge [14, 15].

There are also great risks to indigenous peoples and communities that are documenting illegal extractive or other activities in their territories. Since 2020, Brazilian "Comando Vermelho" assassins for hire by those responsible for drug trafficking and environmental crime in remote Amazon areas have entered much of the eastern Peruvian Amazon, and the situation is similar in most of the other Amazon countries [16]. Over the past decade, and increasingly in the most recent years, at least 33 indigenous leaders and environmental defenders have been assassinated in the Peruvian Amazon alone, and the numbers are higher for Colombia and Brazil [17–19].

6. Other experiences

An important, early analysis of the utility of geospatial services in support of Amazonian indigenous peoples in Peru was published in 2003 [20]. In most cases non-indigenous non-profit organisations and sometimes government agencies employ technical expertise in support of the indigenous peoples' needs, rather than the IPOs directly, although that situation is rapidly changing as increasingly IPOs are acquiring the necessary software and skills using them for this purpose.

The best known process of the use of geospatial services in support of Amazon indigenous territories is that of the Amazon Network of Georeferenced Socio-Environmental Information (RAISG), headquarted in the Brazilian non-profit Instituto Socioambiental (ISA). RAISG is a collective initiative of Amazonian non-profit organisations that include the Instituto do Homen e Meio Ambiente (IMAZON) in Brazil, Fundación Amigos de la Naturaleza (FAN) in Bolivia, Instituto del Bien Común (IBC) in Peru, Fundación Gaia Amazonía in Colombia, and Provita and Wataniba in Venezuela, as well as ISA [21]. Like SERVIR Amazonia, it is supported by international cooperation programs. In some ways RAISG is similar to SERVIR Amazonia in that it fosters access to geospatial information in indigenous territories, although it covers all nine Amazon countries. Also, it prioritises the use of ESRI's ArcGIS, to which it subscribes, rather than freely-accessible no-cost NASA-produced and other programs.

An Australian source emphasises the greater sense of place that indigenous peoples have and encourages better communication between them and geospatial scientists, so that the latter may better understand the territories that they are mapping [22].

7. Conclusions

By collectively and directly integrating remote sensing, analysis, and mapping into land-use and development planning in territories historically occupied by indigenous peoples, geospatial scientists can improve their understanding of the data that they generate and also empower the indigenous peoples to more effectively manage and defend those territories. Initiatives, like SERVIR Amazonia can demonstrate the power of such cooperative efforts not only by preventing deforestation and environmental degradation, but also foster community resilience through their direct participation in the decision-making that affects their territories and their future lives.

Engagement of indigenous peoples in highly technical geospatial services is not an easy task, but it is an important one for the programs like SERVIR Amazonia that seek to help defend highly biodiverse indigenous forest lands against increasing threats coming from external actors. It requires well-informed indigenous organisations that understand and assume the issues involved and also institutions and individuals that are seriously committed to working with indigenous peoples and addressing their territorial problems. The trainers must be sensitive to cultural differences, willing to use simple, easily understood language that people lacking extensive technical training can understand. They must be willing to treat indigenous participants as equals and with sympathy, as well as master the complexities of the technical subject matter.

The SERVIR Amazonia Strategy for the Engagement of Indigenous Peoples through their representative organisations is highly appropriate. However, for locally-specific services, it is easier to work with the more locally-based indigenous federations than to expect effective coordination from regional or national indigenous organisations based in the capital cities.

Moreover, funding for critical infrastructure and computers is necessary and not likely to be provided by the indigenous partner organisations. SERVIR Amazonia is intended to be a network for spreading geospatial services, not a funding source for indigenous organisations, but to effectively engage and include indigenous peoples and their territories, the necessary budgetary and organisational resources must be found somewhere.

Among the most important personal characteristics for trainers and others who work directly with indigenous leaders and community members are great cultural sensitivity, a willingness to treat indigenous counterparts as equals whose opinions and perspectives matter. They also need strong communication skills that avoid technical jargon; speak plainly, clearly, and simply; in order not just to be understood, but also to listen carefully and fully. There must be an awareness that indigenous peoples know a lot more about their territories than the western geospatial scientists that are studying and mapping them do, and that they have a lot to learn from that. Communication must be a two-way street.

Ethical considerations, such as data ownership, stewardship, and sharing, among others, must be taken into consideration. Programs that provide geospatial services to indigenous peoples and communities must respect the need for privacy; not all of their territorial data should or can be made public. Finally, there must be an awareness of the risks that indigenous leaders and environmental defenders are taking, and their territorial data must be protected against illegal drug traffickers, gold miners, timber harvesters, and traffickers in both land and persons.

Acknowledgements

The author would like to acknowledge the support from SERVIR Amazonia and, in particular, of its Director, Carlos Gasco, in engaging indigenous peoples. He also recognises the technical and social skills demonstrated by SERVIR Amazonia partners, such as the University of Richmond's Geography program, ACCA, ECOCIENCIA, Instituto Sinchi, the Upper Amazon Conservancy, and, the participating indigenous organisations, particularly FENAMAD, ACONADIYSH, OPIRJ, and CONFENIAE. Helpful explanations of the technological issues involved were offered by Glenn Hyman, Brian Zutta, and Waldir Azaña. Funding for the program was provided by USAID and NASA. However, responsibility for the content of this chapter is the author's.

Conflict of interest

The sole author declares no conflicts of interest.

Author details

Thomas Moore
International Centre for Tropical Agriculture (CIAT) Lima, Peru

*Address all correspondence to: tm1854@gmail.com

References

[1] Moore T. Including indigenous peoples in geospatial services. Frontiers in Climate. 2022;**4**:810428. Available from https://www.frontiersin.org/articles/10.3389/fclim.2022.810428/full

[2] SERVIR Amazonia webpage. 2024. Available from: https://servir.alliancebioversityciat.org/#

[3] SERVIR Amazonia, Strategy for the Engagement of Indigenous Peoples. 2019. Available from: https://www.academia.edu/113099282/SERVIR_Amazonia_Strategy_for_Engagement_of_Indigenous_Peoples

[4] IWGIA. The Indigenous World, 2020. Copenhagen: IWGIA; 2021. Available from: https://iwgia.org/en/resources/indigenous-world.html

[5] Aikhenvald AY. Languages of the Amazon. Oxford: Oxford University Press; 2012. ISBN: 9780198723011

[6] Epps P, Salanova AP. The languages of Amazonia, Tipití. Journal of the Society for the Anthropology of Lowland South America. 2013;**11**(1):1-28. Available from: https://digitalcommons.trinity.edu/tipiti/vol11/iss1/1

[7] ICCA Consortium. 2021 Report. Territories of Life. Available from: https://report.territoriesoflife.org

[8] Walker WS, Gorelik SR, Baccini A, Aragon Osejo JL, Josse C, Meyer C, et al. The role of forest conversion, degradation, and disturbance in the carbon dynamics of Amazon indigenous territories and protected areas. Publications of the National Academy of Science (PNAS). 117;**6**:3015-3025. Available from: https://www.pnas.org/doi/epdf/10.1073/pnas.1913321117

[9] Qin Y, Xiao X, Liu F, Sa Silva EF, Shimabokuro Y, Arai E, et al. Forest conservation in indigenous territories and protected areas in the Brazilian Amazon. Nature Sustainability. 2023;**6**:295-305. DOI: 10.1038/s41893-022-01018-z

[10] Fa JE, Watson JEM, Leiper I, Potapov P, Evans TD, Burgess ND, et al. The importance of indigenous peoples' lands for the conservation of intact forest landscapes. Frontiers in Ecology and the Environment. 2020;**18**(3):135-140. Available from https://www.jstor.org/stable/26986195

[11] Hoorn R. Indigenous Land Rights Key to Curbing Deforestation and Restoring Lands: Study. Menlo Park, California: Mongabay LATAM; 2023. Available from: https://news.mongabay.com/2023/06/indigenous-land-rights-key-to-curbing-deforestation-and-restoring-lands-study/

[12] Finer M, Mamani N. Protected Areas and Indigenous Territories Effective against Deforestation across Amazon. No. 183. Washington: Monitoring of the Andean Amazon (MAAP); 2023. Available from: https://www.maaproject.org/2023/protected-indigenous-amazon/

[13] Harttner J, Ryan SJ, MacKenzie C, Parker JN, Strasser CA. Spatially explicit data: Stewardship and ethical challenges in science. PLoS Biology. 2013;**11**:e1001634. DOI: 10.1371/journal.pbio.1001634

[14] Sletto B. Indigenous people don't have boundaries: Reborderings, fire management, and productions of authenticities in indigenous landscapes. Cultural Geography. 2009;**16**(2):253-277. DOI: 10.1177/1474474008101514

[15] Borrero RM. Indigenous Peoples and the Information Society: Emerging Uses of ICTs. Paris: UNESCO Program Document; 2016. Available from: https://unesdoc.unesco.org/ark:/48223/pf0000260781#

[16] Huerta Bustamante P, Brehaut I. Comando Vermelho, banda de narcotraficantes brasileña, se arraiga en el Perú. 2023. Available from: https://data.larepublica.pe/narcotrafico-en-peru-banda-brasilera-comando-vermelho-se-arraiga-en-la-amazonia-ucayali-frontera-brasil-comando-rojo-pnp/#:~:text=El%20Comando%20Vermelho%20es%20una,el%20acrónimo%20CV%20infunde%20terror

[17] Peru Support Group. Kichwa leader killed in San Martín. 2023. Available from: https://perusupportgroup.org.uk/2023/12/kichwa-leader-killed-in-san-martin

[18] Peru Support Group. Controversy surrounds the death of Kakataibo indigenous defender. 2024. Available from: https://perusupportgroup.org.uk/2024/01/controversy-surrounds-the-death-of-kakataibo-indigenous-defender/

[19] Global Witness. Standing firm: the land and environmental defenders on the front lines in the climate crisis. 2023. Available from https://www.globalwitness.org/en/campaigns/environmental-activists/standing-firm/

[20] Smith RC, Benavides M, Pariona M, Tuesta E. Mapping the past and future: Geomatics and territories in the Peruvian Amazon. Human Organization. 2003;**62**(4):357-368. Available from: https://www.jstor.org/4: stable/44127816

[21] RAISG. About RAISG. 2023. Available from: https://www.raisg.org/en/about/

[22] Briggs C, Burfurd I, Duckham M, Guntarik O, McMillan M, Saldías SM, et al. Bridging the geospatial gap: Data about space and indigenous knowledge of place. Geography Compass. 2020;**14**:11. DOI: 10.1111/gec3.12542